Smart Polymeric Nanocomposites: Synthesis and Applications

Edited by

Partha Pratim Saikia
Department of Chemistry
Nanda Nath Saikia College
Titabar, Assam
India

&

Bhaskar Jyoti Saikia
Department of Chemistry
Nanda Nath Saikia College
Titabar, Assam
India

Smart Polymeric Nanocomposites: Synthesis and Applications

Editors: Partha Pratim Saikia and Bhaskar Jyoti Saikia

ISBN (Online): 978-981-5322-04-0

ISBN (Print): 978-981-5322-05-7

ISBN (Paperback): 978-981-5322-06-4

First published in 2025.

Bentham Science Publishers Pte. Ltd.
80 Robinson Road #02-00
Singapore 068898
Singapore
Email: subscriptions@benthamscience.net

BENTHAM SCIENCE

CONTENTS

FOREWORD

Considering the importance of smart polymeric nanocomposites, this book covers a wide variety of chapters covering various aspects of this growing field. This is a huge asset for researchers particularly those who are working in the field of smart polymeric materials. Those who wish to initiate their research in this field can consult this book as a reference as it gives a glimpse of the latest developments in the field of nanocomposites. I hope this book will be beneficial for students of higher education as a part of their curriculum. Covering many aspects of smart polymeric nanocomposites starting from their synthesis, characterization, and applications in various fields such as self-healing, shape memory, drug delivery, water purification, flexible electronic sensing, *etc* it equips readers with a holistic understanding of smart polymeric nanocomposites.

With contributions from esteemed researchers and experts in the field, "**Smart Polymeric Nanocomposites: Synthesis and Applications**" offers a comprehensive overview of the latest advancements and future directions in this exciting area of study.

I hope this book serves as a valuable resource for researchers, academics, and industry professionals alike, inspiring further exploration and innovation in the field of smart polymeric nanocomposites.

I wish all the success in the attempt made by the editors and authors.

Swapan Kumar Dolui
Department of Chemistry
Tezpur University
Tezpur, India

PREFACE

Smart polymeric nanocomposites have attracted the attention of both scientists and researchers for the last few decades as they confer the materials with virtuous processability and sensitivity towards many external stimuli such as light, temperature, pH, electrolyte concentration, magnetic field, *etc*. In the dynamic landscape of materials science, the field of smart polymeric nanocomposites stands at the forefront of innovation, offering a spectrum of possibilities that outdo traditional material boundaries. These smart polymer nanocomposites find applicability in a wide range of applications such as sensors and actuators, drug delivery, self-healing structural material, smart textiles, smart electronics, aerospace applications, *etc*.

This book provides a broad updated survey and information on major categories of smart polymeric nanocomposites such as self-repairing materials, shape-memory materials, stimuli-responsive materials, *etc*. Synthetic routs and applications of all these materials are discussed. The essence of this book is centered on its dedication to encouraging inventive methods in tailoring polymeric nanocomposites to meet the characteristic requirements of cutting-edge applications. From the fields of biomedicine and wastewater treatment to the forefront of self-healing materials, food packaging, sensing technologies, and drug delivery systems, the book reveals a wide range of opportunities where intelligent polymeric materials take on a crucial role.

This book is a culmination of the collective expertise of scholars and experts from around the globe, exploring the fundamental principles that strengthen the design and utilization of smart polymeric nanocomposites. We are highly grateful to all contributing authors/ coauthors who contributed to this book and made our journey successful. We are also grateful to the Bentham Science Publication and its support team for helping us during the completion of this book.

Partha Pratim Saikia
Department of Chemistry
Nanda Nath Saikia College
Titabar, Assam
India

&

Bhaskar Jyoti Saikia
Department of Chemistry
Nanda Nath Saikia College
Titabar, Assam
India

List of Contributors

Amlan Puzari	Department of Chemistry, Nanda Nath Saikia College, Titabar-785630, Assam, India
Arup Kumar Dutta	Department of Chemistry, Pandit Deendayal Upadhyaya Adarsha Mahavidyalaya, Behali, Assam 784184, India
Bhaskar Jyoti Saikia	Department of Chemistry, Nanda Nath Saikia College, Titabar, Assam, India
Basanta Saikia	Department of Chemistry, KU Leuven, Leuven-3001, Belgium
Champak Dutta	Department of Chemistry, Bahona College, Assam-785101, India
Geeti Kaberi Dutta	Department of Chemistry, Bahona College, Bahona, Jorhat-785101, Assam, India
Jayanta Madhab Borah	Department of Chemistry, Nanda Nath Saikia College, Titabar-785630, Assam, India
Kiranjyoti Mohan	Department of Chemistry, Rangapara College, Sonitpur-784505, Assam, India
Khanindram Baruah	Department of Chemistry, North-Eastern Hill University, Shillong, Meghalaya, India
Moon Mandal	Department of Chemistry, Pandit Deendayal Upadhyaya Adarsha Mahavidyalaya, Dalgaon, Darrang-784116, Assam, India
Nimisha Kashyap	Department of Chemistry, North-Eastern Hill University, Shillong, Meghalaya, India
Nobomi Borah	Department of Chemical Sciences, Tezpur University, Tezpur-784028, Assam, India
Partha Pratim Saikia	Department of Chemistry, Nanda Nath Saikia College, Titabar, Assam, India
Pronob Gogoi	Department of Chemistry, Diphu Government College, Diphu, Karbi Anglong, Assam, India
Prithiviraj Khakhlary	Department of Chemistry, Dibrugarh University, Dibrugarh, Assam, India
Rajesh Mandal	Department of Chemistry, Pandit Deendayal Upadhyaya Adarsha Mahavidyalaya, Behali, Assam 784184, India
Rajkumar Gogoi	Department of Physics and Astronomy, University of Manchester, Manchester, United Kingdom
Tumpa Paul	Department of Chemistry, Darrang College, Tezpur-784001, India
Zakaria Halim	Department of Chemical Engineering, Golaghat Engineering College, Bogorijeng, Golaghat-785621, Assam, India

CHAPTER 1

Smart Polymer Nanocomposites: An Overview

Partha Pratim Saikia[1] and **Bhaskar Jyoti Saikia**[1,*]

[1] *Department of Chemistry, Nanda Nath Saikia College, Titabar, Assam, India*

Abstract: Smart polymer nanocomposites have gained a wide range of applications in diverse fields of science and technology as they confer the material's great processability and sensitivity to a wide range of environmental stimuli, including magnetic fields, light, temperature, pH, and electrolyte concentration. Incorporating functional or pristine nanoparticles/nanofillers into polymer matrices significantly modifies the properties of smart polymers, including their shape memory ability, self-healing ability, pH sensitivity, thermal responsiveness, electro responsiveness, *etc*. These smart polymer nanocomposites find applicability in a wide range of applications such as sensors and actuators, drug delivery, self-healing structural material, smart textiles, smart electronics, aerospace applications, *etc*. This chapter seeks to address the importance, synthetic strategies, characterization techniques, and application of smart polymeric nanocomposites, summarizing an overview of the range of these materials.

Keywords: Actuators, Diels-Alder reaction, Graphene, *In-situ* polymerization, Melt-processing, Nanoparticles, Nnanocomposites, Nanotubes, Pi-Pi interaction, Solution-blending, Self-assembly, Smart polymers, Stimuli-responsive polymers.

INTRODUCTION

Smart materials, which are also called stimuli-responsive materials, have gained considerable interest among the scientific community in the last few decades owing to their unique properties of sensitivity towards many external stimuli such as light, temperature, pH, electrolyte concentration, magnetic field, *etc*. Polymers are the most extensively used materials for developing stimuli-responsive materials as their chemistry allows easy modification of the properties by integrating functional chemical moieties. On the other hand, there are many polymers such as Poly(N-isopropyl acrylamide), Poly(acrylic acid), Poly(ethylene glycol), Poly(2-hydroxyethyl methacrylate), Polyethyleneimine, *etc*., which themselves are responsive towards external stimulus [1 - 4]. These polymers can detect one or more stimuli as a signal, determine the extent of the change, and then directly respond to the stimulus by changing the conformation of their

* **Corresponding author Bhaskar Jyoti Saikia:** Department of Chemistry, Nanda Nath Saikia College, Titabar, Assam, India; Tel: +91-8486202290; +91-7002497058; E-mail: bhaskarsaikia222@gmail.com

chains. These responses show up as alterations in the solubility, volume, shape, degree of association, and other properties [5, 6].

Polymer nanocomposites are materials that combine organic and inorganic components with at least one filler dimension less than 100 nm. Due to the increased interfacial area between the polymers and the nano-fillers, they exhibit better mechanical and physical characteristics than the host polymers [7 - 9]. Incorporation of pristine or functional nanoparticles into a polymer matrix can drastically enhance and modify the properties due to their high surface area, reinforcing effects, nucleating effects, and intrinsic functionalities. These nanomaterials can be derived from special polymers such as shape memory polymers, stimuli-responsive polymers, electro-responsive polymers, self-repairing polymers, self-cleaning polymers, sensingpolymers, *etc.* directly or by adding responsive nanoparticles. Resulting smart nanocomposites have a wide range of applications such as sensors and actuators [10], drug delivery [11], self-heling structural material [12], smart textiles [13], smart electronics [14], aerospace applications [15], *etc.*

In the last few decades, numerous efforts have been made for the integration of properties of nanoparticles and polymers for designing stimuli-responsive materials. Hence, a large number of organic and inorganic nanofillers have been studied that can amplify or generate stimuli-responsive behaviour. Based on their responsive behaviours, nanofillers can be classified into two categories [16 - 18]. The first category of nanofillers is called inert nanofillers, which have no stimuli-responsive behavior. These include cellulose, silica, and clay materials. The second type includes carbon nanotubes, graphene, metal oxide nanoparticles, *etc.*, which show stimuli-responsive behaviour along with reinforcement. These can be termed active nanofillers. For the synthesis of smart polymer nanocomposites, the second type of nanofillers is primarily used due to their ability to respond to different stimuli. These nanofillers interact with polymer matrix mostly through weak intermolecular force of attraction. Due to the high surface area of nanofillers, which favors agglomeration, a good distribution of nanoparticles in the polymer matrix is difficult to achieve. Hence, fabrication techniques of polymer nanocomposites have a profound effect on the final composite. Different synthetic strategies are used for this purpose including, melt processing [19], solution blending [20], *in-situ* polymerization methods [21], *etc.*, which are discussed in detail in this chapter.

The aim of this chapter is to give an overview of various smart polymer nanocomposites showcasing diverse responsive behaviors such as shape memory effect, self-healing characteristics, pH responsiveness, and electro responsiveness.

Moreover, this chapter will focus on different synthetic strategies, characterization techniques, and utilization of these materials in various applications.

SMART POLYMER NANOCOMPOSITES

Shape Memory Polymer Nanocomposites

Biocompatible progression in polymer properties has led to the development of materials capable of adapting their physicochemical properties and/or structural conformations in response to the surrounding environment, analogous to those found in natural organisms known as "smart polymers" or "stimuli-responsive polymers". Shape-changing polymers are a class of stimuli-responsive polymers that undergo molecular or visible-level shape alterations [22]. In contrast, shape memory polymers (SMPs), another type of polymer, retain their original shape after a brief applied force is removed. The dual-shape memory polymer effect is the change in the shape of SMPs from a permanent to a temporary state upon application of a transient applied force [23]. The three stages of strain that shape memory polymers (SMPs) go through are: (i) applying strain above the glass transition temperature; (ii) deforming to the desired shape close to room temperature; and (iii) removing the force such that the polymer returns to its initial shape above the glass transition temperature [24]. The active controllable temperature of SMPs can be sustained through changes in composition and degree of cross-linking.

Various reinforcements are incorporated into the SMPs to enhance the recoverable strain. According to studies, adding nanostructures to polymers greatly enhances their interface characteristics, while the strength and modulus are increased by the microstructure [25, 26]. Consequently, the utilization of nanofillers like nanoparticles, nanotubes, and nanowires in SMP nanocomposites is gaining importance. For instance, the addition of nano-sized SiO_2 to polylactide-based polyester enhances recovery temperature, flexibility, and mechanical strength, making it suitable for biomedical applications [27]. Likewise, the incorporation of carbon nanofibers into epoxy-based shape memory nanocomposites improves Young's modulus and product strength [28]. There have been several studies published in which SMP nanocomposites are fabricated on resins using a variety of methods, including resin transfer molding (RTM), vacuum-assisted resin transfer molding (VARTM), filament winding, and resin film infusion (RFM). In the following section, different nanofillers used for the reinforcement of SMPs will be discussed.

Graphene

Diffusion-controlled graphene-based nanomaterials have been widely used as nanofillers because of their effectiveness in creating shape-memory polymer nanocomposites with exceptional shape memory and improved glass transition temperature. The incorporation of graphene oxide and reduced graphene oxide as reinforcement into a poly(vinyl alcohol) polymer matrix has resulted in an increase in glass transition temperature (Tg) significantly, from 81.10 ^{0}C to 106.5 ^{0}C and 101.15 ^{0}C respectively [29]. The increase in the glass transition temperature is explained by a strong interfacial contact between nanofillers and the polymer matrix. Because of its enhanced hardness and improved crystallinity, the nanocomposite showed exceptional mechanical capabilities, including shape recovery and shape fixity values of up to 95%. Another study found that the addition of graphene nanoplatelets (GNPs) to an SMP nanocomposite consisting of polycaprolactone, thermoplastic starch, and graphene showed sufficient healing capabilities, with 90% shape fixity and 70% shape recovery [30]. The healing capacity is further enhanced with the strategic positioning of GNPs within the starch phase; however, excessive GNPs led to agglomeration and reduced performance. Furthermore, the shape memory ability under water stimulus can be increased in the presence of glycerol. In another work, a 3D epoxy polymer nanocomposite was fabricated with graphene nanoplatelets resulting in a near-set shape SMPs [31]. The incorporation of GNPs has improved the tensile strength by 30% and elasticity by 17% in the 3D printed composites, along with a noteworthy 365% increase in the storage modulus compared to the pure polymer nanocomposite. Reinforcement of SMPs with combined GO and carbon nanotube (CNT) nanofiller exhibits notable electrical and mechanical properties [32]. While GO deflects polymer fissures in the polymer matrix, CNTs bridge those cracks, increasing interfacial loading and greatly raising the polymer composite's elastic modulus. Moreover, the storage modulus at room temperature increased by 40% due to the effective dispersion of nanofillers, which restricted the chain mobility and also enhanced mechanical properties.

Carbon Nanotubes (CNT)

The utilization of electrical energy for nanocomposite manipulation provides numerous advantages, particularly in achieving a high degree of shape recovery for diverse applications that are electrostimulated. Fabricating SMPs through electrical actuation provides several benefits for creating high-performance electroactive systems. 3D polymer networks incorporated with CNTs enhance mechanical properties, with storage modulus increasing from 3.0 GPa to 2.4 GPa and the flexural stress during shape recovery reaches 14.6 MPa [33]. The addition of CNTs also elevates the glass transition temperature (Tg) from 90°C to 100°C enabling the polymer nanocomposite to respond within 8 seconds when voltage is applied, due to the random charge distribution within the 3D network. Polyhydroxyethylmethacrylate (PHEMA) reinforced with multi-walled carbon nanotubes (MWCNTs) demonstrated electrical conductivity and remote healing proficiencies with a heat-triggered healing response [34]. It was observed that an increase in the MWCNT content improves Tg and thermomechanical properties, leading to improved electroactive properties with reduced voltage requirements. Furthermore, by using the Joule effect, electrical heating increases share fixity and recovery, reaching maximum levels of 100% and 95%, respectively. Both thermally and electrically induced recovery is demonstrated by hybrid MWCNTs and SMPs with montmorillonite nano clay incorporation [35]. However, due to the greater impact of nanofillers on thermal actuation, electrically induced recovery exhibits a faster trend. Additionally, the nanofiller dramatically reduces recovery time at a pace 10 times quicker than heat activation. Thermally-induced shape memory is made possible by the incorporation of halloysite nanotubes (HNTs) and MWCNTs into a polyurethane composite to create a 4D-printed polymer nanocomposite [36]. HNTs in combination with MWCNTs enhance the polymer healing by 34%, resulting in a high flexural strength increase of approximately 57%. The resulting polymer achieves a significant recovery of shape memory in just 120 seconds, with higher MWCNT content further enhancing recovery ability. Moreover, the nanofillers boost the crystallinity of the polymer by acting as nucleating agents, thereby enhancing cross-linking density and elevating Tg.

Other Fillers

Besides graphene and CNTs, other nanomaterials are also gaining importance as reinforcement to the polymer matrix. It has been noted that adding magnetic nanoparticles to nanofillers improves their capacity to regain form and gives the polymer matrix superparamagnetic characteristics. For instance, Fe_3O_4-infused bio-based benzoxazine resin induces healing ability triggered by magnetic field and light [37]. The nanofiller infusion resulted in increased fixity from 85% to 93%, with polymer shape recovery attaining 98% and magnetic field-induced recovery reaching 99% within 26 seconds, making it suitable for soft robotics applications. For SMPs to be successful, the concentration of the nanomaterial must be optimized since too much of it may hinder form recovery. For instance, when concentration increases, silica nanoparticles can decrease both for shape fixity and self-healing [38]. On the other hand, the addition of silica nanoparticles to polymer matrices increases the mechanical strength by 100% fixity *via* multifunctional crosslinking. It was shown in a different investigation that adding SiO_2 nanoparticles to mixtures of hard and soft polymers greatly increases fixity and recovery values. Polydopamine nanofillers reinforced in a PCL/TPU blend have been shown to display light-actuated recovery, achieving 100% shape recovery and healing efficiency of 78.5% within 150 seconds [39]. It was further observed that increased PCL content enhances shape recovery and shape-fixing ability. Additionally, the polymer blend's shape memory was improved with the addition of PDA. It has been demonstrated that adding cellulose CNTs to polymeric materials can give them the capacity to sense water [40]. A flexible photonic polymeric material with remarkable mechanical qualities and multi-cycle repeatability is created by combining PU with disulfide bonds and CNTs. The film gains toughness *via* the S-S bond, which allows it to support up to 5500 times its weight.

Self-Healing Polymer Nanocomposites

Self-healing polymers (SHP) represent another class of stimuli-responsive polymers that have gained recognition for their versatile applications. SHPs possess the ability to autonomously recover their properties following damage or corrosion. These include various materials such as elastomers, thermosets, supramolecular polymers, nanocomposite polymer coatings, and polymer composites, which have been categorized into autonomous and non-autonomous types. By adding various nanomaterials, these polymers' mechanical, electrical, and magnetic characteristics may be adjusted in response to various stimuli [41 - 44]. They find utility in various biomedical applications like tissue engineering, wound healing, antibacterial properties, *etc*.

The properties of functional polymers, such as toughness, Young's modulus, strength, *etc*., can be significantly enhanced by the addition of fillers and reinforcements like clay, cellulose, metals, and carbonaceous materials. The healing ability of polymer nanocomposites can be stimulated through various methods, including the Joule effect, in which an electric current passes through a nanostructured network and causes thermal healing [45]. Both intrinsic and extrinsic features characterize self-healing materials. Block copolymers with reactive functional groups that undergo reversible bond interactions in response to environmental stimuli constitute the intrinsic system [46, 47]. In contrast, extrinsic properties offer limited healing capacity beyond a stimulus threshold [48, 49]. Self-healing occurs on its own in response to external factors, offering advantages for a variety of applications, such as energy, coatings, electronics, and biomedical devices. Crosslinking, chain movement, physical connection, polymerization, and chemical interactions all contribute to the self-healing process in polymer matrixes [50 - 52]. Subsequently, quite a few factors that stimulate the healing include chemical repair, chain entanglement, and polymerization. The incorporation of nanostructures enhances the polymer properties, which facilitates the development of high-performance multipurpose materials. However, creating high-performing, multifunctional, self-healing nanocomposites requires achieving a homogenous dispersion of nanoparticles. Moreover, self-healing characteristics are influenced by variables like morphology, nanoparticle content, and microphase separation.

Methods of Self-Healing in Self-Healing Polymer Nanocomposites

Diels-Alder Reaction

The Diels-Alder (DA) reaction is the reaction of a diene and a dienophile to form cyclohexene, which can also be reversed through a retro Diels-Alder reaction, which results in the breakdown of products into the starting materials. Both these reactions are controlled by temperature, where higher temperature favors the Diels-Alder reaction while lower temperature favors the retro Diels-Alder reaction [53]. The sensitivity of polymer nanocomposites to temperature affects their structure, improving their mechanical and self-healing properties. Additionally, different nanofillers can be utilized in the polymer matrix to facilitate both the Diels-Alder and retro Diels-Alder reactions.

Multi-Walled Carbon Nanotubes (MWCNTs)

In the healing mechanism involving the Diels-Alder reaction, MWCNTs serve as both diene and dienophile, easing the DA reaction. The self-healing and mechanical qualities of furfuryl-linked styrene-butadiene rubber are enhanced by including furfuryl-terminated MWCNTs (MWCNT-FA) [54]. MWCNT-FA serves as both a healing agent and reinforcement, facilitating DA reaction with maleimide and furan moieties. The polymer nanocomposites that are produced exhibit temperature responsiveness, as maleimide experiences the DA reaction when the temperature drops from 160°C to 74°C and the rDA reaction when the temperature drops even lower to 80°C. The addition of magnetic nanoparticles imparts magnetic behavior to the composite, serving as a stimulant for healing when a magnetic field is present. The addition of a magnetic copolymer results in supermagnetic behavior when it is exposed to a magnetic field and vanishes when the magnetic field is removed. Carbon nanotubes coated with polycaprolactone-poly(furfuryl glycidyl ether) and iron oxide nanoparticles exhibit self-healing through the rDA mechanism. Here, the DA reaction takes place among the dienophile of polycaprolactone-poly(furfuryl glycidyl ether) and diene of MWCNT, producing the DA adducts that exhibit supermagnetic behavior [55]. The incorporation of MWCNTs with a furan-derived polymer composite exhibits a self-healing mechanism through a three-dimensional network through DA reaction. The method shows its ability for hydrogen bond-induced healing as well as DA/rDA healing. When CNTs are directly added to furan pendant polymers, they exhibit active, temperature-dependent behavior and a substantial healing ability in response to thermal and electrical stimuli. Defect sites in carbon have been found to play a critical role in the DA reaction, offering more active sites for a greater cross-linked mesh inside the polymer matrix.

Graphene Derivatives

Functional graphene oxides have been discovered to be an effective filler in self-healing polymer nanocomposites, alongside carbon nanotubes. The healing ability of these materials is mostly based on the Diels-Alder/retro-Diels-Alder process. To create self-healing nanocomposites, Lin and coworkers synthesized furan as pendant groups and terminal maleimide groups for crosslinking in polyurethane [56]. Reduced graphene oxide (rGO) was utilized as a filler and silver nanowires (AgNWs) as conductive materials, which improved healing potential *via* the DA reaction mechanism. AgNWs and rGO worked together to offer the polymer remarkable near-infrared-induced healing capabilities. It was discovered that the size of the rGO particles affected the nanocomposites' mechanical and self-healing characteristics. Moreover, an increase in the content of functional graphene oxide reduces the mechanical efficiency due to the decreased flexibility of the polymer chain. The incorporation of nanocomposite polymers with furfurylamine-modified diglycidyl ether of bisphenol A and furfuryl functionalized aniline trimer-modified graphene with bismaleimide (BMI) exhibited high tensile strength and electrical conductivity [57]. However, it has been found that if the graphene is poorly disseminated, there is less interfacial contact, which lowers the polymer qualities.

Quadruple Hydrogen Bonding

The self-healing ability and mechanical properties of nanocomposites, including thermoresponsive and crystalline supramolecules, can be enhanced by hydrogen bonding (HB), allowing the polymer to be customized for enhanced toughness and consistency. Materials that are rich in hydrogen bonding gradually recover their electrostatic interactions, ensuring high healing ability. One of the most common hydrogen bonding contributors is Ureidopurimidinone (UPy), which is introduced into the polymer *via* the reaction of 2-amino-4-hydroxy- 6-methylpyrimidine (AHMP) with isocyanate [58]. A supramolecular polymer nanocomposite was synthesized by Lin and co-workers by combining SiO_2 functionalized with UPy and polycarbonatediol end-capped with UPy [59]. UPy-decorated polyurethane has been found to exhibit self-healing characteristics at room temperature, with the healing efficiency dependent on UPy concentration. The quadruple hydrogen bonding *via* interaction between SiO_2-UPy and UPy has significantly increased Young's modulus and tensile strength. The increase in UPy content restricts intermolecular cross-linking, permitting enhanced chain mobility and stronger quadruple hydrogen bonding, thereby increasing healing ability. The incorporation of magnetic Fe_3O_4 nanoparticles into poly(n-butyl acrylate) was found to impart superparamagnetic properties [60]. Additionally, quadruple hydrogen bonding inside the polymer matrix is facilitated by the incorporation of 2-ureido-4[1H]-pyrimidinone methyl methacrylate during copolymerization, which enhances the polymer's thermochemical characteristics and ability for self-healing.

Pi-Pi Interaction

The mechanical properties of polymeric materials can be enhanced by increasing p-p interactions, although attaining complete healing in these interactions normally requires prolonged exposure to high temperatures [61, 62]. For instance, Vajyapuri and co-workers have revealed that optimization of nanoparticle content promotes better p-p interactions, leading to enhanced mechanical qualities and healing effectiveness [63]. They have reported the synthesis of a shape memory polymer nanocomposite having polyamide functionalized with pyrene and gold nanoparticles functionalized with both pyrene and polyamide, which shows p-p stacked interactions.

Metal-Ligand Co-Ordination

Metal-ligand coordination polymers are those polymers that have a central metal ion surrounded by polymers and offer optical and stimuli-responsive properties. The dynamic non-covalent bonds between the polymer matrix and metal account for the stiffness and strength of these composites [64]. Moderate bond energy in metal-ligand interactions is responsible for association and dissociation, enhancing the healing ability. For example, a polyurethane ligand terminated with terpyridine polymerized on multi-walled carbon nanotubes cross-linked by Zn^{2+} ions displays strength, toughness, and elasticity [65]. The complexation-decomplexation mechanism is what drives the healing process.

Types of Stimuli

Temperature

Thermoresponsive nanocomposites are frequently used as smart materials because the temperature is a convenient external stimulus. In response to temperature variations, they undergo structural changes associated with hydrogen bonds. The most typical thermoresponsive polymers are hydrogels, which demonstrate a reversible volume phase transition from a hydrated state to a collapsed state, displaying significant volume changes. The transition temperature is dependent on whether it occurs upon heating or cooling: when it decreases, *e.g.*, in poly(acrylamide) and poly(acrylic acid), it is termed as upper critical solution temperature [66, 67], whereas, in the case the transition occurs on heating, it is known as lower critical solution temperature as observed in poly(N-vinyl caprolactum), poly(oxazoline), polyethylene oxide, *etc* [68].

Magnetic Field

Magnetic nanoparticles respond to static or alternating magnetic fields, which arise due to the interaction between magnetic moments in the material and the magnetic gradient by the field [69]. Owing to the ease of synthesis, superparamagnetic properties, and biocompatibility, metals, alloys, and metal oxides are commonly used as magnetic nanoparticles [70]. Polymers served as the matrices for supporting and stabilizing well-dispersed nanomagnetic particles to form magnetic polymer nanocomposites. These polymeric nanocomposites respond to the external magnetic field, resulting in a change in shape or movement, making them suitable for different applications such as purification and drug delivery systems. In addition, magnetic nanomaterials have been shown to exhibit thermal effects, known as hyperthermia, under alternating magnetic fields. This finds applications in various fields, including actuators, shape-memory, cancer treatment, and drug delivery.

Electric Field

The application of electric fields is interesting in shape-memory polymer or shape-memory polymer nanocomposite, as it can produce a change in the shape of the material depending on the release of local heat produced through the Joule heating effect. It is important to synthesize the polymer nanocomposite with conductive matrices or fillers. Initially, conducting polymers were extensively studied, which require reinforcement with inert fillers to enhance their mechanical and thermal properties [71 - 74]. These days, carbon-based nanofillers such as graphene, graphene oxide, and carbon nanotubes are being studied because of their superior mechanical, thermal, and electrical characteristics, even if their randomized distribution has certain drawbacks [75]. Efforts have also been made to combine the properties of conductive polymers and nanofillers. Yang *et al.* have synthesized an electrically controllable nanocomposite containing 2-acrylamide-2-methyl propane sulfonic acid and acrylamide with reduced graphene oxide as a filler having swelling/deswelling behavior [76]. The polymer nanocomposite was found to display enhanced mechanical and electrical properties compared to the pure polymer matrix, with faster response time making it suitable for soft robotics applications.

Light

Due to the ease of generation, light is an attractive stimulus in various scientific fields. In contrast to electric or magnetic fields, light serves as a remote-control stimulation force unaffected by electromagnetic interference. When interacting with light, nanomaterials, particularly metals or metal oxide exhibit localized surface plasmon resonances, a property determined by their nature, shape, and size [77]. When light strikes an intense electromagnetic field, it creates a fast heat transfer from the nanoparticles to the surrounding environment. Light-responsive polymer nanocomposites find widespread application in medicine, particularly in drug delivery and cancer treatment. Due to its ability in deep tissue penetration and minimal absorbance in human tissue compared to UV light, near-infrared irradiation is favored for medical applications. For example, Li and co-workers developed a polyethyleneimine matrix modified with thiocarbamate as a platform for near-infrared activated H_2S release [78]. Exogenous H_2S has shown promising anti-cancer activity by inhibiting tumor growth.

Overview of Synthesis Methods for Smart Polymer Nanocomposites

Smart polymeric materials combine the characteristics of polymers and nanoparticles, resulting in materials with improved functionality such as self-healing, shape memory, actuation, and sensing. Preparation of these polymer nanocomposites involves the dispersion of functional nanoparticles in the matrices of polymer to achieve desired properties. Preparation of superior polymer nanocomposites using an appropriate processing technique is crucial for achieving high-quality polymer nanocomposites. Different synthetic methods aim to incorporate nanoparticles into the polymer matrix in a controlled manner. An overview of some frequently used methods is discussed below:

Solution Blending

Solution blending is the easiest and the most economical method for the preparation of polymer nanocomposites, as it requires no complex arrangements and a large number of chemicals. This method relies on evaporating solvents from the solutions containing the nano dispersion. The first step for solution blending is polymer and nanoparticle selection. Smart polymers and corresponding nanoparticles with specific properties have to be selected depending on the desired properties of the final polymer. These polymers included polymethyl methacrylate, polyethylene, polypropylene, alkyl resins, epoxy resins, *etc*. For imparting smart behavior to the final product, sometimes it is necessary to modify nanoparticles through surface modifications. Various surface modification techniques have been used, including chemical functionalization, polymer coating, core sell formation, oxide layer formation, doping, surface patterning, *etc*

[79]. Solvent selection plays an important role in the synthesis of nanocomposites. The solvent has to be compatible with both the polymer and nanoparticles; in the meantime, it should have good drying properties. Common solvents include tetrahydrofuran, chloroform, dimethylformamide, and toluene. To achieve uniform dispersion, nanoparticles have to be dispersed in suitable solvents using techniques such as mechanical mixing, high-shear mixing, ultrasonication, *etc.* Nanoparticle dispersion and polymer solutions are then mixed together while stirring continuously. This step allows the nanoparticles to be evenly distributed within the polymer matrix. The final step for the synthesis of polymer nanocomposites using the solution blending method is solvent evaporation. This is achieved using techniques such as rotary evaporation, vacuum drying, *etc.* to get the final product.

Several studies have demonstrated the use of solution-blending methods for the preparation of smart polymer nanocomposites. The manufacture of poly(vinyl chloride) (PVC) filled with micro- and nanosized calcium carbonate ($CaCO_3$) *via* a solution-blending approach was described by Gulsen Albayrak Ari *et al.* [80]. In this report, PVC and $CaCO_3$ nanoparticles were premixed and dry-blended. A solution of PVC-$CaCO_3$ was produced in tetrahydrofuran (THF) and was stirred with magnetic stirring for about 30 min until the solution became homogenous. The resulting solution was heated under reflux, and stirring was continued for 210 min. Finally, solvent evaporation was carried out to get the polymer nanocomposites. Numerous reports are available on the synthesis of carbon nanotube-polymer nanocomposites *via* solution blending [81]. Functionalized CNTs, especially those having polymer grafts, result in better dispersion compared to pristine CNTs in a polymer solution, thereby forming superior nanocomposites.

Melt Mixing

Melt blending, or melt mixing, is another widely used method for the synthesis of smart polymer nanocomposites that involves the controlled addition of nanoparticles to a molten polymer. Like any method for the synthesis of polymer nanocomposites, the first step is selection of polymer and nanoparticle with desired properties. The next step is pre-mixing. For the nanoparticles to properly disperse in the polymer matrix, this pre-mixing procedure is carried out. In the pre-mixing step, the desired smart polymer and corresponding nanoparticles are mixed using suitable equipment, including twin-screw extruders, internal mixers (such as a Banbury mixer), or kneaders. This blend is then heated above the melting point of the polymer under vigorous mixing to obtain a homogeneous nanocomposite. Melt excluders or similar equipment are used for the purpose.

Finally, the melt is extruded into a cold chamber or cooled surface. This solidifies the melt, and the final smart polymer nanocomposite is obtained.

The melt-mixing method offers the benefit of scalability, cost-effectiveness, versatility, and being environmentally friendly due to the non-involvement of any volatile organic solvents and hence, if permissible, a much better alternative for solution mixing. Superior mechanical, thermal, and functional properties are obtained due to enhanced interfacial interaction between components synthesized by this method. Despite all these advantages, poor temperature control limits applications of this method for the synthesis of polymer nanocomposites. For instance, Junior *et al.* [82] synthesized high-impact organoclay/polystyrene (PS) nanocomposites using melt blending. An interpenetrating corotating twin screw extruder was used for the procedure. Similarly, Maiti *et al.* [83] reported on the production of extremely low loading of MWCNTs *via* melt-blending of polycarbonate and styrene acrylonitrile–MWCNT, resulting in highly electrically conductive polycarbonate (PC)/styrene acrylonitrile (SAN)/multi-wall carbon nanotubes (MWCNTs) nanocomposites. They observed that following the melt-mixing of PC and SAN-MWCNT, well-dispersed MWCNTs formed in the SAN, providing a continuous, conductive interconnected network structure throughout the matrix.

In Situ Polymerization Methods

In situ polymerization has numerous advantages over other methods for the synthesis of polymer nanocomposites. Both thermoplastic and thermosetting plastics can be blended with nanoparticles using this method. Moreover, it allows the grafting of polymers on the surfaces of nanoparticles, offering numerous benefits. This method involves the polymerization of suitable monomers in the presence of desired nanoparticles. The process follows two major steps. In the first step, nanoparticles are dispersed within the monomer solution and uniformly dispersed before polymerization. Surface modification or functionalization of nanoparticles may be employed to improve compatibility with the monomer and promote uniform dispersion. In the next step, a suitable initiator, crosslinker, or catalyst is added, and favorable polymerization conditions are applied. As the polymerization proceeds, it forms a network of polymers with embedded nanoparticles. By regulating the degree of polymerization, the size and morphology of the polymer matrix can be tailored. Finally, a suitable purification process is carried out to remove unused monomers, catalysts, and other impurities.

Fe_3O_4/poly (ε-caprolactone)-polyurethane (PCLU) shape memory nanocomposites were synthesized by Yan Cai *et al.* [84] using an *in-situ* polymerization technique.

It was observed that the Fe_3O_4 nanoparticles were evenly distributed inside the PCLU matrix, ensuring the nanocomposites' excellent shape memory properties in both hot water and an alternating magnetic field. By adjusting the temperature to around 34 °C, Yan Deng *et al.* [85] demonstrated the reversible self-assembly and disassembly of graphene oxide functionalization, enabling the creation of thermo-sensitive nanocomposites using *in situ* SET-LRP. Tris(hydroxymethyl) aminomethane was used to initiate an epoxide ring-opening reaction on exfoliated GO sheets since GO does not contain any functional groups. The initiating groups were attached to Tris(hydroxymethyl) aminomethane-GO sheets by reacting hydroxyls with 2-bromo-2-methylpropionyl bromide.

This technique enjoys many advantages, like uniform nanoparticle distribution, tailored properties, enhanced interfacial interaction, scalability, and reproducibility. It holds more potential when nanoparticles are functionalized. Although this technique shows potential in producing polymer nanocomposites with consistent and distinct architectures, it is suitable for small-scale production only, as very small amounts of the final product are obtained.

Other Methods

Apart from the aforementioned techniques, researchers investigated different methods along with modified conventional techniques for better dispersion of nanoparticles in a polymer matrix for achieving superior properties after fabrication. For instance, the *in-situ* polymerization method can be modified to a photo-induced, catalytic, redox, or even chain transfer process. Other such approaches include one-pot synthesis, microwave-induced synthesis, electrochemical synthesis, template-directed synthesis, inter-matrix synthesis (IMS), self-assembly synthesis, *etc* [86].

IMPORTANCE OF CHARACTERIZATION IN EVALUATING SMART POLYMER NANOCOMPOSITES

Knowledge of characterization techniques plays a crucial role in understanding the structure, performance, and properties of smart polymers. These properties, including physical, chemical, morphological, mechanical, and thermal properties, determine the overall performance of polymer nanocomposites. Smart polymeric nanocomposites possess various smart characteristics regarding their ability to respond to external stimuli or dynamic conditions. These responsive behaviors include responses to specific triggers such as temperature, pH, light, electric fields, or mechanical stress. Different characterization techniques are utilized for the evaluation of smart behavior and structure-property relationships of polymer nanocomposites. These techniques provide valuable insights for the design, optimization, and quality control of polymer nanocomposites. In the following section, we discuss how different characterization techniques help the study of smart polymer nanocomposites.

Crystallographic Properties

Improved filler dispersion and interfacial interactions would regulate nanocomposites' performance. The method of X-ray diffraction is crucial for optimizing such types of interactions. It is the most commonly used approach for evaluating the crystalline and amorphous properties of fillers, polymers, and composite materials. It helps to illustrate the impact of nanoparticles on the polymer's crystalline behavior and phase composition. Selected Area Electron Diffraction (SAED) is another characterization technique that provides detailed information about the crystallographic structure and orientation of nanoparticles within the nanocomposite material. A transmission electron microscope is used for SAED. In contrast, another method for investigating the crystallographic properties of polymer nanocomposites is small-angle X-ray scattering (SAXS). In terms of parameters like averaged particle sizes, chemical composition, and crystal orientation, as well as particle size distribution, morphologies, and surface-to-volume ratio, SAXS is used to determine the microscale or nanoscale structure of polymer nanocomposites. The purpose of this approach is to investigate the structural alterations of nanocomposites that do not exhibit a particular order and do not provide a Bragg diffraction peak. For example, XRD is widely used for characterization of clay-polymer nanocomposites. The XRD spectra of clays display a distinctive diffraction peak owing to their unique layered structure. Surface modification of polymer intercalation brings about changes in d-spacing, peak position, or peak intensity in the XRD pattern of clay. XRD peaks of immiscible polymer and clay show no change in these parameters while intercalation increases the d-spacing. Franci Malin *et al.* studied polyacryl–MMT nanoclay composites for anti-corrosion applications. In MMT clay, the interlayer distance is 1.87 nm, and the peak is located at $2\theta=4.72^{\circ}$. However, in nanocomposites involving Polyacryl, the peak moves to $2\theta=2.44^{\circ}$, indicating an interlayer distance of 3.61 nm and indicating that the polymer chains have reached the clay's interlayer area [87].

Structural Characterization Techniques

Through the use of micro- or nano-level imaging of nanocomposites, morphological changes in polymers following the incorporation of nanoparticles in the matrix are investigated. These techniques enable visualization of nanoparticle dispersion, size, shape, and morphology within the polymer matrix. The commonly assessed morphological characteristics include the dispersion of nanoparticles in the matrix, the change in surface texture, the change in failure characteristics, the distribution and mapping of nanoparticles, and their composition on the surface. Scanning Electron Microscopy (SEM), Transmission Electron Microscopy (TEM), and Atomic Force Microscopy (AFM) are the most widely used techniques for this purpose. SEM and TEM enable the picturing of nanoparticle dispersion, shape, size, and morphology within the polymer matrix, while AFM provides high-resolution surface imaging, which assists with testing the nanoscale morphology of the polymer-nanoparticle to probe the nanoscale morphology of the polymer-nanoparticle interface and measure surface roughness. Combining energy-dispersive X-ray spectroscopy with scanning electron microscopy enables the analysis of nanoscale surfaces and offers insight into the elemental composition. To obtain more detailed and high-resolution images, atomic force microscopy (AFM) or transmission electron microscopy (TEM) is preferred over scanning electron microscopy (SEM) or field emission SEM (FESEM) as AFM and TEM provide atomic-level resolutions. This makes AFM and TEM more appropriate for imaging at the nanometer scale than SEM or FESEM, which typically have lower resolution capabilities, on the order of a few hundred nanometers. For example, Yong Yang *et al.* used an aerosol single-drop reactor technique to show how sub-5 nm Ni, Co, and Cu nanoparticles could be generated within a polymer matrix. Using SEM, the surface morphology of the synthesized metal-polymer nanocomposites comprising Ni, Co, and Cu nanoparticles was examined. The particles were found to have a spherical shape, and their sizes varied roughly from several hundred nanometers to 1.5 micrometres (Fig. **1**) [88].

Fig. (1). SEM image of (**a**) Ni polymer (**b**) Co-polymer (**c**) Cu-polymer nanocomposites (Reprinted with permission from [88]).

Mechanical Characterization Methods

Tensile testing using a universal testing machine, along with dynamic mechanical analysis (DMA), is the main mechanical characterization method for any polymer nanocomposites. Tensile testing includes tensile strength, elastic modulus, and elongation at break, which helps to evaluate the reinforcement effect of nanoparticles on the material's mechanical performance. DMA measures the viscoelastic properties of nanocomposites as a function of temperature, frequency, or strain amplitude. It provides insights into the material's response to mechanical stress and temperature changes. These testing methods are extensively used for determining the healing efficiency of smart self-healing polymers.

Thermal Analysis (DSC, TGA)

The incorporation of nanoparticles into a polymer matrix modifies the thermal behavior of nanocomposites. Nanocomposites display improved thermal properties over pristine polymers due to high nanoparticle-polymer interfacial area. Thermogravimetric analysis (TGA) and Differential Scanning Calorimetry (DSC) are used to study the thermal behavior of polymer nanocomposites, which includes melting point, glass transition temperature, decomposition temperature, *etc*.

APPLICATIONS OF SMART POLYMERIC NANOCOMPOSITES

Owing to their combined properties of both polymer and nanoparticles, polymer nanocomposites have a wide range of applications. Smart polymeric nanocomposites find their applicability across various fields of material science owing to their dynamic, responsive behavior. Since both physical and chemical interactions take place at nanoscales between smart polymers and functional nanoparticles, or *vice versa*, smart polymer nanocomposites demonstrated extraordinary properties and design possibilities. Different applications of smart polymer nanocomposites can be studied under the following categories:

Sensors and Actuators

Stimuli-responsive polymer nanocomposites have tremendous applications in the field of detection of materials. Because of their large surface area, fast electron mobility, high conductivity value, ease of interaction with analyte molecules, quick reaction time, sensitivity, and selectivity, polymer nanocomposites are the most promising materials for sensing. They can be tailored to change their electrical, optical, or mechanical properties with temperature changes, which will create sensors for monitoring applications. Conducting polymer-based nanocomposites are good candidates for temperature sensors. Polyaniline, polythiophene, polypyrrole, and their derivatives are the most common conducting polymers for sensing. For instance, Luheng Wang developed carbon nanotube polyurethane foam composites and carbon black-filled silicon rubber composites as property-induced differential temperature sensors [89].

Chemical sensors possess the ability to detect specific chemicals by investigating changes in the electrical conductivity, optical properties, and different electrochemical properties in the presence of the target analyte. Polymer nanocomposites are extensively used for building chemical sensors owing to their promising features such as excellent electrical conductivity, high electrocatalytic activities, and effective surface area in comparison to pristine polymers. Matlou GG *et al.* prepared the single-walled carbon nanotube poly(m-amino benzenesulfonic acid)-modified gold electrode by the self-assembled monolayers procedure for the detection of heavy metal Hg2+, which exhibited high sensitivity and a limit of detection equal to 0.0634 μM [90]. Polypyrrole-coated SnO_2 nanocomposites were used for the detection of toxic ammonia gas, which showed the sensitivity of ammonia gas in the range (1–10.7 ppm) at room temperature [91]. Similarly, conducting polymer nanocomposites show potential as materials for detecting gases that pose the greatest threat to the environment, including CO, NO_2, NH_3, and H_2S.

Some smart polymeric nanocomposites can undergo measurable changes in capacitance, electrical resistance, or mechanical properties in response to applied mechanical forces and thus can be applied as stress-strain sensors. These kinds of polymer nanocomposites are utilized to monitor changes in mechanical properties brought on by pressure, deformation, or bending. Pedro Costa *et al.* used a styrene-butadiene-styrene thermoplastic elastomer polymer matrix reinforced with carbon nanofillers as conducting material for developing piezoresistive sensors for large strain sensing applications. Additionally, these components were integrated into a hand glove together with the wireless communication system and readout, and it demonstrated that these materials were suitable for tracking finger movements [92]. (Fig. **2**).

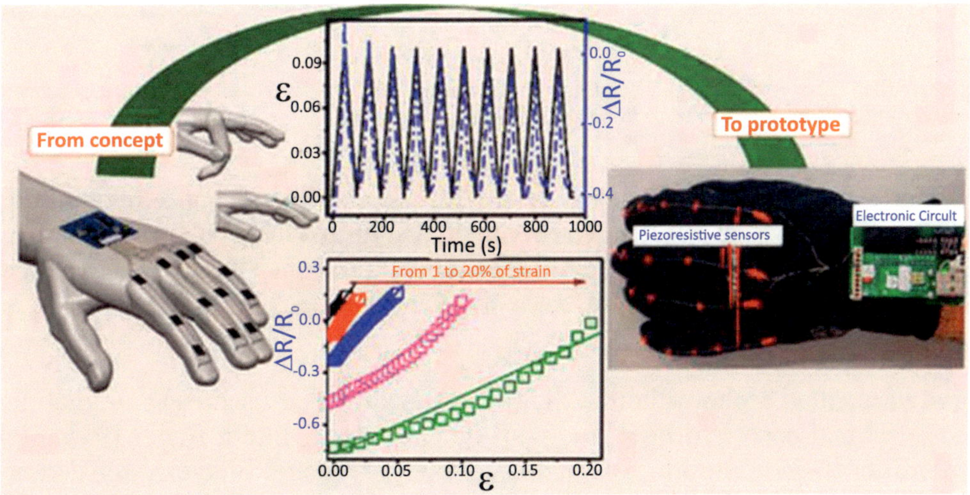

Fig. (2). Strain sensing thermoplastic elastomer polymer matrix reinforced with carbon nanofillers (Reprinted (adapted) with permission from [92]).

Polymer nanocomposites based on shape memory polymers can act as stimuli-responsive actuators that have the ability to change shape in response to external stimuli such as heat, light, microwaves, *etc*. These materials have found their applicability in minimally invasive surgical tools, adaptive structures, morphing aerospace components, *etc*. Because of their entropic elasticity, these polymeric molecular chains exhibit an apparent coil/stretch transition with sliding when energy is consumed by applied stress. The incorporation of functional or pristine nanoparticles, such as graphene oxide, carbon nanotubes, *etc*. can drastically improve the mechanical properties of these shape memory polymers. Graphene oxide-reinforced polyvinyl alcohol nanocomposites with shape memory were developed by Qi *et al*. It was found that the development of more physical cross-linked points between the polyvinyl alcohol and the graphene oxide layer was the reason for the enhanced shape memory characteristics of these nanocomposites [93]. (Fig. **3**).

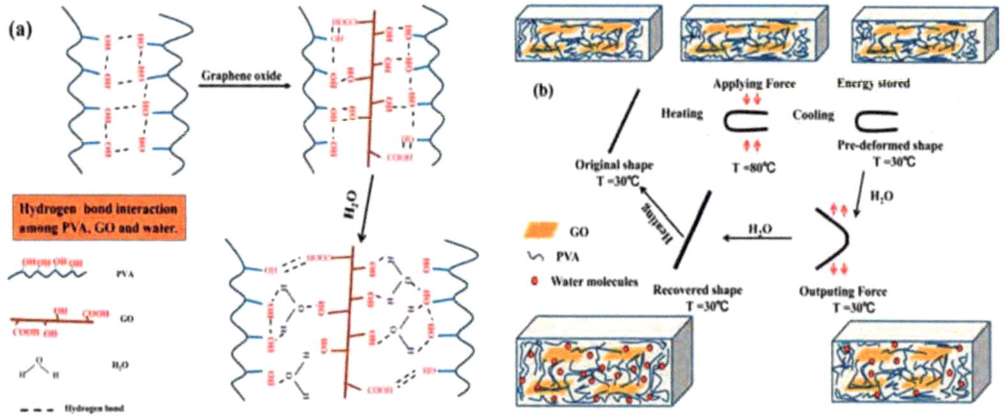

Fig. (3). Illustrations of hydrogen bonding interaction among PVA, GO and water (**a**) and schematic representation of shape memory behaviour (**b**). (Reproduced from [93]).

Polymeric nanocomposites based on hydrogels exhibit changes in volume in response to changes in different environmental stimuli such as pH, temperature, or humidity, which can be converted to mechanical motions. These behaviors make these materials suitable for different applications such as microfluidics, soft robotics, and biomimetic devices.

Shi *et al*. created a photo-thermal hydrogel actuator that used light irradiation to precisely bend each hydrogel fragment one at a time, like a finger [94]. Light-responsive bending motion was achieved using hydrogel bilayers, consisting of nanocomposite poly(N-isopropyl acrylamide). The layers' different expansion coefficients caused the bending motion to happen. In response to local light irradiation, the resulting nanocomposite actuators demonstrated flexible, reversible bending and non-bending actions.

Drug Delivery Systems

Smart polymeric nanocomposites have gained considerable interest among the scientific community due to their stimuli-responsive behavior. These smart materials are suitable for smart drug delivery systems because they undergo transformations in response to biochemical or physical stimuli as well as minute external environmental variations. These polymeric composites can detect one or more stimuli as a signal, determine the strength of the signal, and then react immediately by changing the conformations of their chains. For instance, poly(N-isopropyl acrylamide) is the most studied temperature-responsive polymer, which can undergo reversible phase transformation at about 32 ^{0}C. To generate an on-off drug release profile in response to a gradual change in temperature, these kinds of polymers and associated temperature-responsive nanocomposites are being employed. Mahsa Asgari M *et al.* designed a multi-stimuli-responsive polymer nanocomposite with a core-shell-shell structure consisting of iron oxide nanoparticles as the core, mesoporous silica as the middle shell, and poly(N-isopropyl acrylamide-co-acrylic acid) as an exterior shell [95]. The synthesized nanocomposite exhibits excellent magnetic and temperature-responsive behavior to be used as a smart drug delivery system. In reaction to an external field, electroresponsive polymers (ERP) exhibit bending, shrinking, or swelling. The hydrogel ERP matrix is capable of encasing several medications. The deswelling of the polymer allows encapsulated drugs to be released from a matrix when exposed to applied electric fields. For the transdermal administration of ketoprofen, Kulkarni *et al.* developed an electroresponsive drug delivery system utilizing poly(acrylamide-grafted-xanthan gum) hydrogel [96]. These kinds of methods work well for transdermal drug delivery systems that offer on-demand drug release and are activated by an electric signal [96]. Inflammation-Responsive Smart polymers are another very interesting type of polymer that undergoes polymer chain lysis in specific sites of the body with inflammation. Inflammations are characterized by the formation of free radicals. Hence any polymer chain that undergoes lysis in the presence of free radicals enables the release of drugs at the site of inflammation. Nobuhiko Yui carried out such an experiment using lipid microspheres in degradable matrices of hyaluronic acid crosslinked with polyglycerol polyglycidylether that constitutes excellent inflammation (stimulus)-responsive implantable drug delivery [97]. pH-responsive polymer nanocomposites are another type of smart material that are extensively used for drug delivery. These pH-responsive polymers can be modified to have a specific pKa value for achieving target-specific drug delivery. The transition pH of polyelectrolytes can be tailored by modulating its pKa or pKb through chemical functionalization of the base polymer. These nanocomposites function through distinct mechanisms in acidic and basic mediums. In an acidic medium such as a tumor microenvironment, polymers with acidic functional groups undergo swelling upon protonation. These lead to structural changes in the nanocomposites and subsequent drug delivery. However, in basic mediums such as intestinal environments, nanocomposites with basic functional groups undergo deprotonation, which leads to shrinking the polymer matrix, thereby controlling drug release. Thus, specific drugs can be delivered at specific sites based on the pH of the delivery site.

Self-Healing Materials

Smart polymeric nanocomposites are extensively used for designing industrially important self-healing polymers. These kinds of materials can regain their intrinsic qualities and strength whenever something damaging happens, whether it be externally (such as scratching) or internally (such as fracture development). Dynamic covalent bond formation, encapsulation of monomer/catalyst to polymer matrix, and supramolecular self-assembly are the most frequently adopted strategies for designing self-healing polymers. Their importance in the advancement of materials science and engineering is underlined by their potential to mitigate structural damage, reduce maintenance costs, and improve product longevity. Extrinsic healing polymeric systems rely on vascular networks or capsules as external healing agents. Poly (urea-formaldehyde), polyurethane, poly (melamine-urea formaldehyde), and poly (melamine-formaldehyde) are the most common materials used to make capsule shells. Materials with intrinsic self-healing properties are based on the development of dynamic covalent bonds or non-covalent bonds. Self-healing polymer nanocomposites can be successfully integrated into concrete structures to heal cracks and damage caused by shrinkage, seismic activity, or chemical exposure, thus enhancing durability and longevity. In aerospace applications, materials are exposed to extreme conditions which causes damage to the materials. Self-healing polymer nanocomposites are successfully designed to ensure the structural integrity of these materials. Epoxy composites are the most extensively used structural composites in aerospace. A self-healing epoxy composite containing an incorporated healing ingredient was developed by Teoh *et al.* Indentation was utilized to induce damage, which was subsequently subjected to three-point bend flexural testing. The flexural strength data demonstrated that the self-healing agent inside the hollow glass capillary tubes recovered a substantial portion of the mechanical strength [98]. Similarly, self-healing polymer nanocomposites are extensively used in energy storage industries to improve the reliability and performance of batteries and capacitors by resisting damage-induced degradation. Apart from the aforementioned applications, self-healing polymer nanocomposites are extensively used in packaging industries, textile industries, biomedical devices, automotive industries, *etc.*

Environmental Remediation

Smart polymer nanocomposites offer versatile solutions for environmental remediation challenges, starting from water, soil, and air pollution to hazardous waste management. Polymer nanocomposites have been widely used for the removal of heavy metals from water. Smart adsorbents and filtration membranes based on polymer nanocomposites used in water treatment are responsive to single or multiple stimuli. Functionalized nanoparticles embedded in a polymer matrix possess the ability to selectively adsorb or catalytically degrade pollutants. For example, membranes that are formed from graphene oxide (GO) multilayers and nanobrushes of poly (acrylic acid) show excellent pH-responsive filtration properties. The major characteristics of graphene oxide (GO), such as charge and interlayer spacing, vary dramatically with pH, resulting in various pH-dependent interfacial phenomena and separation processes [99]. For this reason, the GO membrane has considerable potential as a pH-responsive membrane. Another kind of smart filter media that exhibits reversible variations in water permeability produced by magnetic fields is magnetic field-responsive polymer nanocomposites. One such membrane was produced by Himstedt *et al.* by covalently attaching Fe_3O_4 nanoparticles to poly(2-hydroxyethyl methacrylate) chains, which were then grafted onto PET membranes. This membrane exhibited magnetic field-responsive water permeability because of the grafted chains' reversible expansion into the membrane's pores when the magnetic field is on and contraction when it is off [100].

CONCLUSION

The study of smart polymeric nanocomposites has opened up exciting possibilities in materials science and engineering. By sensibly creating and combining appropriate smart polymers and nanoparticles, these materials have shown remarkable responsiveness to external stimuli, leading to a wide range of state-of-the-art applications such as self-healing, pH sensitivity, thermal responsiveness, electro-responsiveness, *etc.* The versatility and intelligence within these nanocomposites hold great potential for enhancing people's lives. But to make these materials useful in real life and profitable, the right measures must be taken toward scaling them up. This chapter summarizes different synthetic strategies, characterization techniques, and applications of smart polymeric nanocomposites. However, polymer nanocomposites require more comprehensive study, since due to the inherent properties of the nanoparticles, realizing their homogeneous dispersion in the polymer matrix has always been a challenge. Hence, studies focusing on new strategies for the better dispersion of nanoparticles in polymer matrix will revolutionize the field of smart polymeric nanocomposites.

REFERENCES

[1] Wei, M.; Gao, Y.; Li, X.; Serpe, M.J. Stimuli-responsive polymers and their applications. *Polym. Chem.,* **2017**, *8*(1), 127-143.
[http://dx.doi.org/10.1039/C6PY01585A]

[2] Liu, F.; Urban, M.W. Recent advances and challenges in designing stimuli-responsive polymers. *Prog. Polym. Sci.,* **2010**, *35*(1-2), 3-23.
[http://dx.doi.org/10.1016/j.progpolymsci.2009.10.002]

[3] Bajpai, A.K.; Bajpai, J.; Saini, R.; Gupta, R. Responsive polymers in biology and technology. *Polym. Rev. (Phila. Pa.),* **2011**, *51*(1), 53-97.
[http://dx.doi.org/10.1080/15583724.2010.537798]

[4] Alarcón, C.H.; Pennadam, S.; Alexander, C. Stimuli responsive polymers for biomedical applications. *Chem. Soc. Rev.,* **2005**, *34*(3), 276-285.
[http://dx.doi.org/10.1039/B406727D] [PMID: 15726163]

[5] Hu, J.; Liu, S. Responsive polymers for detection and sensing applications: current status and future developments. *Macromolecules,* **2010**, *43*(20), 8315-8330.
[http://dx.doi.org/10.1021/ma1005815]

[6] Schattling, P.; Jochum, F.D.; Theato, P. Multi-stimuli responsive polymers – the all-in-one talents. *Polym. Chem.,* **2014**, *5*(1), 25-36.
[http://dx.doi.org/10.1039/C3PY00880K]

[7] Fu, S.; Sun, Z.; Huang, P.; Li, Y.; Hu, N. Some basic aspects of polymer nanocomposites: A critical review. *Nano Materials Science,* **2019**, *1*(1), 2-30.
[http://dx.doi.org/10.1016/j.nanoms.2019.02.006]

[8] Krishnamoorti, R.; Vaia, R.A. Polymer nanocomposites. *J. Polym. Sci., B, Polym. Phys.,* **2007**, *45*(24), 3252-3256.
[http://dx.doi.org/10.1002/polb.21319]

[9] Tjong, S.C. Structural and mechanical properties of polymer nanocomposites. *Mater. Sci. Eng. Rep.,* **2006**, *53*(3-4), 73-197.
[http://dx.doi.org/10.1016/j.mser.2006.06.001]

[10] Rivero, R.E.; Molina, M.A.; Rivarola, C.R.; Barbero, C.A. Pressure and microwave sensors/actuators based on smart hydrogel/conductive polymer nanocomposite. *Sens. Actuators B Chem.,* **2014**, *190*, 270-278.
[http://dx.doi.org/10.1016/j.snb.2013.08.054]

[11] Mohammadpour-Haratbar, A.; Zare, Y.; Rhee, K.Y. A review on drug delivery systems containing polymer nanocomposites for breast cancer treatment. *Polym. Rev. (Phila. Pa.),* **2023**, 1-38.

[12] Guadagno, L.; Vertuccio, L.; Barra, G.; Naddeo, C.; Sorrentino, A.; Lavorgna, M.; Raimondo, M.; Calabrese, E. Eco-friendly polymer nanocomposites designed for self-healing applications. *Polymer (Guildf.),* **2021**, *223*, 123718.
[http://dx.doi.org/10.1016/j.polymer.2021.123718]

[13] Ghosh, S.; Nitin, B.; Remanan, S.; Bhattacharjee, Y.; Ghorai, A.; Dey, T.; Das, T.K.; Das, N.C. A multifunctional smart textile derived from merino wool/nylon polymer nanocomposites as next generation microwave absorber and soft touch sensor. *ACS Appl. Mater. Interfaces,* **2020**, *12*(15), 17988-18001.
[http://dx.doi.org/10.1021/acsami.0c02566] [PMID: 32207964]

[14] Gong, M.; Zhang, L.; Wan, P. Polymer nanocomposite meshes for flexible electronic devices. *Prog.*

Polym. Sci., **2020**, *107*, 101279.
[http://dx.doi.org/10.1016/j.progpolymsci.2020.101279]

[15] Njuguna, J.; Pielichowski, K.; Fan, J. *Polymer nanocomposites for aerospace applications*; Advances in Polymer Nanocomposites, **2012**, pp. 472-539.

[16] Njuguna, J.; Ansari, F.; Sachse, S.; Rodriguez, V.M.; Siqqique, S.; Zhu, H. *Nanomaterials, nanofillers, and nanocomposites: types and properties. InHealth and environmental safety of nanomaterials*; Woodhead Publishing, **2021**, pp. 3-37.

[17] Jayawickrama, MT; Ishra, MI; Nishadani, BS; Manthilaka, MM Outline and classification of inorganic nanofillers in polymers. *InNanofillers,* , 1-23.
[http://dx.doi.org/10.1201/9781003279389-1]

[18] Sharma, A.K. *Priya, Kaith BS. Polymer nanocomposite matrices: classification, synthesis methods, and applications. InHandbook of Polymer and Ceramic Nanotechnology*; Springer International Publishing: Cham, **2021**, pp. 403-428.

[19] Wang, L.; Gardner, D.J.; Wang, J.; Yang, Y.; Tekinalp, H.L.; Tajvidi, M.; Li, K.; Zhao, X.; Neivandt, D.J.; Han, Y.; Ozcan, S.; Anderson, J. Towards industrial-scale production of cellulose nanocomposites using melt processing: A critical review on structure-processing-property relationships. *Compos., Part B Eng.,* **2020**, *201*, 108297.
[http://dx.doi.org/10.1016/j.compositesb.2020.108297]

[20] Madaleno, L.; Schjødt-Thomsen, J.; Pinto, J.C. Morphology, thermal and mechanical properties of PVC/MMT nanocomposites prepared by solution blending and solution blending+melt compounding. *Compos. Sci. Technol.,* **2010**, *70*(5), 804-814.
[http://dx.doi.org/10.1016/j.compscitech.2010.01.016]

[21] Gohari, D.P.; Kalaee, M.R.; Sharif, A. Interfacial *in situ* polymerization of layered-silicate/poly (hexamethylene isophthalamide) nanocomposites. *J. Inorg. Organomet. Polym. Mater.,* **2019**, *29*(4), 1243-1251.
[http://dx.doi.org/10.1007/s10904-019-01088-1]

[22] Lendlein, A.; Kelch, S. Shape-Memory Polymers. *Angew. Chem. Int. Ed.,* **2002**, *41*(12), 2034.
[http://dx.doi.org/10.1002/1521-3773(20020617)41:12<2034::AID-ANIE2034>3.0.CO;2-M] [PMID: 19746597]

[23] Yang, Z.; Herd, G.A.; Clarke, S.M.; Tajbakhsh, A.R.; Terentjev, E.M.; Huck, W.T.S. Thermal and UV shape shifting of surface topography. *J. Am. Chem. Soc.,* **2006**, *128*(4), 1074-1075.
[http://dx.doi.org/10.1021/ja056866s] [PMID: 16433513]

[24] Gall, K.; Dunn, M.L.; Liu, Y.; Finch, D.; Lake, M.; Munshi, N.A. Shape memory polymer nanocomposites. *Acta Mater.,* **2002**, *50*(20), 5115-5126.
[http://dx.doi.org/10.1016/S1359-6454(02)00368-3]

[25] Bhattacharya, S.K.; Tummala, R.R. Epoxy nanocomposite capacitors for application as MCM-L compatible integral passives. *J. Electron. Packag.,* **2002**, *124*(1), 1-6.
[http://dx.doi.org/10.1115/1.1400751]

[26] Ash, B.J.; Rogers, D.F.; Wiegand, C.J.; Schadler, L.S.; Siegel, R.W.; Benicewicz, B.C.; Apple, T. Mechanical properties of Al_2O_3/polymethylmethacrylate nanocomposites. *Polym. Compos.,* **2002**, *23*(6), 1014-1025.
[http://dx.doi.org/10.1002/pc.10497]

[27] Yan, B.; Gu, S.; Zhang, Y. Polylactide-based thermoplastic shape memory polymer nanocomposites. *Eur. Polym. J.,* **2013**, *49*(2), 366-378.
[http://dx.doi.org/10.1016/j.eurpolymj.2012.09.026]

[28] Dong, Y.; Ni, Q.Q.; Li, L.; Fu, Y. Novel vapor-grown carbon nanofiber/epoxy shape memory nanocomposites prepared *via* latex technology. *Mater. Lett.,* **2014**, *132*, 206-209.
[http://dx.doi.org/10.1016/j.matlet.2014.06.084]

[29] Canbay, C.A.; Ünlü, N. Production and characterization of shape memory polymeric nanocomposite materials. *J. Mol. Struct.,* **2021**, *1227*, 129708. [http://dx.doi.org/10.1016/j.molstruc.2020.129708]

[30] Shahsavari, E.; Ghasemi, I.; Karrabi, M.; Azizi, H. Starch/polycaprolactone/graphene nanocomposites: shape memory behavior. *Iran. Polym. J.,* **2023**, *32*(6), 763-772. [http://dx.doi.org/10.1007/s13726-023-01166-9]

[31] Idowu, A.; Thomas, T.; Boesl, B.; Agarwal, A. Cryo-assisted extrusion three-dimensional printing of shape memory polymer–graphene composites. *J. Manuf. Sci. Eng.,* **2023**, *145*, 041003.

[32] Roman, J.; Zhao, X.; Whitney, C.; Venkatesan, K.R.; Dai, L.L.; Chattopadhyay, A. High-performance multifunctional shape memory epoxy with hybrid graphene oxide and carbon nanotube reinforcement. *J. Mater. Eng. Perform.,* **2023**, 1-11.

[33] Peng, Q.; Wei, H.; Qin, Y.; Lin, Z.; Zhao, X.; Xu, F.; Leng, J.; He, X.; Cao, A.; Li, Y. Shape-memory polymer nanocomposites with a 3D conductive network for bidirectional actuation and locomotion application. *Nanoscale,* **2016**, *8*(42), 18042-18049. [http://dx.doi.org/10.1039/C6NR06515E] [PMID: 27740656]

[34] Cortés, A.; Cosola, A.; Sangermano, M.; Campo, M.; González Prolongo, S.; Pirri, C.F.; Jiménez-Suárez, A.; Chiappone, A. DLP 4D-printing of remotely, modularly, and selectively controllable shape memory polymer nanocomposites embedding carbon nanotubes. *Adv. Funct. Mater.,* **2021**, *31*(50), 2106774. [http://dx.doi.org/10.1002/adfm.202106774]

[35] Mat Yazik, M.H.; Hameed Sultan, M.T.; Jawaid, M.; Mazlan, N.; Abu Talib, A.R.; Md Shah, A.U.; Safri, S.N.A. Shape memory properties of epoxy with hybrid multi-walled carbon nanotube and montmorillonite nanoclay nanofiller. *Polym. Bull.,* **2024**, *81*(1), 951-968. [http://dx.doi.org/10.1007/s00289-023-04750-4]

[36] Namathoti, S.; Vakkalagadda, M.R.K. Development of multiwalled carbon nanotubes/halloysite nanotubes reinforced thermal responsive shape memory polymer nanocomposites for enhanced mechanical and shape recovery characteristics in 4D printing Applications. *Polymers (Basel),* **2023**, *15*(6), 1371. [http://dx.doi.org/10.3390/polym15061371] [PMID: 36987152]

[37] Leungpuangkaew, S.; Amornkitbamrung, L.; Phetnoi, N.; Sapcharoenkun, C.; Jubsilp, C.; Ekgasit, S.; Rimdusit, S. Magneticand light-responsive shape memory polymer nanocomposites from bio-based benzoxazine resin and iron oxide nanoparticles. *J. Mater. Eng. Perform.,* **2023**, *6*, 215-225.

[38] Choong, Y.Y.C.; Maleksaeedi, S.; Eng, H.; Yu, S.; Wei, J.; Su, P.C. High speed 4D printing of shape memory polymers with nanosilica. *Appl. Mater. Today,* **2020**, *18*, 100515. [http://dx.doi.org/10.1016/j.apmt.2019.100515]

[39] Chen, Y.; Zhao, X.; Luo, C.; Shao, Y.; Yang, M.B.; Yin, B. A facile fabrication of shape memory polymer nanocomposites with fast light-response and self-healing performance. *Compos., Part A Appl. Sci. Manuf.,* **2020**, *135*, 105931. [http://dx.doi.org/10.1016/j.compositesa.2020.105931]

[40] Xue, R.; Zhao, H.; An, Z.W.; Wu, W.; Jiang, Y.; Li, P.; Huang, C.X.; Shi, D.; Li, R.K.Y.; Hu, G.H.; Wang, S.F. Self-healable, solvent response cellulose nanocrystal/waterborne polyurethane nanocomposites with encryption capability. *ACS Nano,* **2023**, *17*(6), 5653-5662. [http://dx.doi.org/10.1021/acsnano.2c11809] [PMID: 36897210]

[41] Thakur, V.K.; Kessler, M.R. Self-healing polymer nanocomposite materials: A review. *Polymer (Guildf.),* **2015**, *69*, 369-383. [http://dx.doi.org/10.1016/j.polymer.2015.04.086]

[42] Billiet, S.; Hillewaere, X.K.D.; Teixeira, R.F.A.; Du Prez, F.E. Chemistry of crosslinking processes for self-healing polymers. *Macromol. Rapid Commun.,* **2013**, *34*(4), 290-309.

[http://dx.doi.org/10.1002/marc.201200689] [PMID: 23255325]

[43] Zwaag, S. An introduction to material design principles: Damage prevention *versus* damage management. *Self Healing Materials*; Springer: Berlin/Heidelberg, Germany, **2007**, pp. 1-18. [http://dx.doi.org/10.1007/978-1-4020-6250-6_1]

[44] Michal, B.T.; Jaye, C.A.; Spencer, E.J.; Rowan, S.J. Inherently photohealable and thermal shape-memory polydisulfide networks. *ACS Macro Lett.*, **2013**, *2*(8), 694-699. [http://dx.doi.org/10.1021/mz400318m] [PMID: 35606954]

[45] Toncelli, C. Thermoreversibility in Polymeric Systems.*Self-Healing at the Nanoscale*; Amendol, V.; Meneghetti, M., Eds.; CRC Press: Boca Raton, FL, USA, **2011**.

[46] Orellana, J.; Moreno-Villoslada, I.; Bose, R.K.; Picchioni, F.; Flores, M.E.; Araya-Hermosilla, R. Self-healing polymer nanocomposite materials by Joule effect. *Polymers (Basel),* **2021**, *13*(4), 649. [http://dx.doi.org/10.3390/polym13040649] [PMID: 33671610]

[47] Wool, R.P. Self-healing materials: a review. *Soft Matter,* **2008**, *4*(3), 400-418. [http://dx.doi.org/10.1039/b711716g] [PMID: 32907199]

[48] Mohammadmahdi, M.; Maryam, G.; Masoud, M. 2—Basics of self-healing composite materials.*Woodhead Publishing Series in Composites Science and Engineering*; Woodhead Publishing: Sawston, UK, **2020**, pp. 15-31.

[49] Zhang, Z.P.; Rong, M.Z.; Zhang, M.Q. Self-healable functional polymers and polymer-based composites. *Prog. Polym. Sci.,* **2023**, *144*, 101724. [http://dx.doi.org/10.1016/j.progpolymsci.2023.101724]

[50] Kausar, A.; Ahmad, I.; Maaza, M.; Bocchetta, P. Self-healing nanocomposites—advancements and aerospace applications. *Journal of Composites Science,* **2023**, *7*(4), 148. [http://dx.doi.org/10.3390/jcs7040148]

[51] Das, M.; Parathodika, A.R.; Maji, P.; Naskar, K. Dynamic chemistry: The next generation platform for various elastomers and their mechanical properties with self-healing performance. *Eur. Polym. J.,* **2023**, *186*, 111844. [http://dx.doi.org/10.1016/j.eurpolymj.2023.111844]

[52] Wang, Y.; Shu, R.; Zhang, X. Strong, supertough and self-healing biomimetic layered nanocomposites enabled by reversible interfacial polymer chain sliding. *Angew. Chem. Int. Ed.,* **2023**, *62*(23), e202303446. [http://dx.doi.org/10.1002/anie.202303446] [PMID: 37010017]

[53] Kuhl, N.; Bode, S.; Hager, M.D.; Schubert, U.S. Self-healing polymers based on reversible covalent bonds.*Self-Healing Materials*; Springer: Berlin/Heidelberg, Germany, **2015**, pp. 1-58. [http://dx.doi.org/10.1007/12_2015_336]

[54] Kuang, X.; Liu, G.; Dong, X.; Wang, D. Enhancement of mechanical and self-healing performance in multiwall carbon nanotube/rubber composites *via* diels–alder bonding. *Macromol. Mater. Eng.,* **2016**, *301*(5), 535-541. [http://dx.doi.org/10.1002/mame.201500425]

[55] Lee, Y.H.; Zhuang, Y.N.; Wang, H.T.; Wei, M.F.; Ko, W.C.; Chang, W.J.; Way, T.F.; Rwei, S.P. Fabrication of Self-Healable Magnetic Nanocomposites *via* Diels−Alder Click Chemistry. *Appl. Sci. (Basel),* **2019**, *9*(3), 506. [http://dx.doi.org/10.3390/app9030506]

[56] Lin, C.; Sheng, D.; Liu, X.; Xu, S.; Ji, F.; Dong, L.; Zhou, Y.; Yang, Y. NIR induced self-healing electrical conductivity polyurethane/graphene nanocomposites based on Diels−Alder reaction. *Polymer (Guildf.),* **2018**, *140*, 150-157. [http://dx.doi.org/10.1016/j.polymer.2018.02.036]

[57] Lin, C.; Sheng, D.; Liu, X.; Xu, S.; Ji, F.; Dong, L.; Zhou, Y.; Yang, Y. Effect of different sizes of graphene on Diels-Alder self-healing polyurethane. *Polymer (Guildf.),* **2019**, *182*, 121822.

[http://dx.doi.org/10.1016/j.polymer.2019.121822]

[58] Pan, Y.; Hu, J.; Yang, Z.; Tan, L. From fragile plastic to room-temperature self-healing elastomer: Tuning quadruple hydrogen bonding interaction through one-pot synthesis. *ACS Appl. Polym. Mater.*, **2019**, *1*(3), 425-436.
[http://dx.doi.org/10.1021/acsapm.8b00153]

[59] Lin, F.; Wang, R.; Liu, L.; Li, B.; Ouyang, L.W.; Liu, W.J. Enhanced intermolecular forces in supramolecular polymer nanocomposites. *Express Polym. Lett.*, **2017**, *11*(9), 690-703.
[http://dx.doi.org/10.3144/expresspolymlett.2017.67]

[60] Xu, S.; Zhao, B.; Adeel, M.; Mei, H.; Li, L.; Zheng, S. Shape memory and self-healing properties of polymer-grafted Fe3O4 nanocomposites implemented with supramolecular quadruple hydrogen bonds. *Polymer (Guildf.)*, **2019**, *172*, 404-414.
[http://dx.doi.org/10.1016/j.polymer.2019.04.020]

[61] Orts, W.J.; Shey, J.; Imam, S.H.; Glenn, G.M.; Guttman, M.E.; Revol, J-F. Application of cellulose microfibrils in polymer nanocomposites. *J. Polym. Environ.*, **2005**, *13*(4), 301-306.
[http://dx.doi.org/10.1007/s10924-005-5514-3]

[62] Favier, V.; Chanzy, H.; Cavaille, J.Y. Polymer nanocomposites reinforced by cellulose whiskers. *Macromolecules*, **1995**, *28*(18), 6365-6367.
[http://dx.doi.org/10.1021/ma00122a053]

[63] Vaiyapuri, R.; Greenland, B.W.; Colquhoun, H.M.; Elliott, J.M.; Hayes, W. Molecular recognition between functionalized gold nanoparticles and healable, supramolecular polymer blends – a route to property enhancement. *Polym. Chem.*, **2013**, *4*(18), 4902-4909.
[http://dx.doi.org/10.1039/c3py00086a]

[64] Pooshidani, Y.; Ghofrani, R.; Shabani, I. Nanostructured self-healing polymers and composites.*Fundamentals of Nanoparticles*; Elsevier: Amsterdam, The Netherlands, **2018**, pp. 401-423.
[http://dx.doi.org/10.1016/B978-0-323-51255-8.00013-6]

[65] Zheng, Q.; Ma, Z.; Gong, S. Multi-stimuli-responsive self-healing metallo-supramolecular polymer nanocomposites. *J. Mater. Chem. A Mater. Energy Sustain.*, **2016**, *4*(9), 3324-3334.
[http://dx.doi.org/10.1039/C5TA10694J]

[66] Stamenkovic, V.R.; Mun, B.S.; Arenz, M.; Mayrhofer, K.J.J.; Lucas, C.A.; Wang, G.; Ross, P.N.; Markovic, N.M. Trends in electrocatalysis on extended and nanoscale Pt-bimetallic alloy surfaces. *Nat. Mater.*, **2007**, *6*(3), 241-247.
[http://dx.doi.org/10.1038/nmat1840] [PMID: 17310139]

[67] Fane, A.G.; Wang, R.; Hu, M.X. Synthetic membranes for water purification: status and future. *Angew. Chem. Int. Ed.*, **2015**, *54*(11), 3368-3386.
[http://dx.doi.org/10.1002/anie.201409783] [PMID: 25613795]

[68] Bai, H.; Sun, D.D. Hybrid TiO2 photocatalytic oxidation and ultrafiltration for humic acid removal and membrane fouling control. *Water Sci. Technol. Water Supply*, **2011**, *11*(3), 324-332.
[http://dx.doi.org/10.2166/ws.2011.048]

[69] Wang, Y.; Zhao, M.; Hou, C.; Yang, X.; Li, Z.; Meng, Q.; Liang, C. Graphene-based magnetic metal organic framework nanocomposite for sensitive colorimetric detection and facile degradation of phenol. *J. Taiwan Inst. Chem. Eng.*, **2019**, *102*, 312-320.
[http://dx.doi.org/10.1016/j.jtice.2019.06.019]

[70] Brillas, E.; Sirés, I.; Oturan, M.A. Electro-Fenton process and related electrochemical technologies based on Fenton's reaction chemistry. *Chem. Rev.*, **2009**, *109*(12), 6570-6631.
[http://dx.doi.org/10.1021/cr900136g] [PMID: 19839579]

[71] He, G.; Wen, Y.; Ma, C.; Li, X.; Wang, L.; Gao, L.; Sun, Z. Nanoporous CoFe2O4 Loaded with Pt-Ag for Photocatalytic Hydrogen Evolution. *J. Miner. Met. Mater. Soc.*, **2021**, *73*(9), 2798-2807.

[http://dx.doi.org/10.1007/s11837-021-04762-3]

[72] Qi, K.; Yu, J. Modification of ZnO-based photocatalysts for enhanced photocatalytic activity.*Interface Science and Technology*; Yu, J.; Jaroniec, M.; Jiang, C., Eds.; Elsevier: Amsterdam, The Netherlands, **2020**, Vol. 31, pp. 265-284.

[73] Jin, Z.; Gong, H.; Li, H. Visible-light-driven two dimensional metal-organic framework modified manganese cadmium sulfide for efficient photocatalytic hydrogen evolution. *J. Colloid Interface Sci.,* **2021**, *603*, 344-355.
[http://dx.doi.org/10.1016/j.jcis.2021.06.111] [PMID: 34197984]

[74] Zhang, J.; Yang, L.; Cui, Y.; Tang, Y.L.; Xu, J.Y.; Su, Z.C.; Lao, X.Z.; Bao, Y.T.; Wang, X.R.; Xu, S.J. The role of luminescence self-absorption in photocatalytic properties of self-assembled ZnS nanocrystals. *AIP Adv.,* **2020**, *10*(10), 105314.
[http://dx.doi.org/10.1063/5.0020856]

[75] Liu, X.; Li, J.; Yao, W. CdS@MoS $_2$ hetero-structured nanocomposites are highly effective photocatalysts for organic dye degradation. *ACS Omega,* **2020**, *5*(42), 27463-27469.
[http://dx.doi.org/10.1021/acsomega.0c03968] [PMID: 33134709]

[76] Bhirud, A.; Chaudhari, N.; Nikam, L.; Sonawane, R.; Patil, K.; Baeg, J.O.; Kale, B. Surfactant tunable hierarchical nanostructures of CdIn2S4 and their photohydrogen production under solar light. *Int. J. Hydrogen Energy,* **2011**, *36*(18), 11628-11639.
[http://dx.doi.org/10.1016/j.ijhydene.2011.06.061]

[77] Poizot, P.; Baudrin, E.; Laruelle, S.; Dupont, L.; Touboul, M.; Tarascon, J.M. Low temperature synthesis and electrochemical performance of crystallized FeVO4·1.1H2O. *Solid State Ion.,* **2000**, *138*(1-2), 31-40.
[http://dx.doi.org/10.1016/S0167-2738(00)00784-0]

[78] Wang, H.; Cui, L.F.; Yang, Y.; Sanchez Casalongue, H.; Robinson, J.T.; Liang, Y.; Cui, Y.; Dai, H. Mn3O4-graphene hybrid as a high-capacity anode material for lithium ion batteries. *J. Am. Chem. Soc.,* **2010**, *132*(40), 13978-13980.
[http://dx.doi.org/10.1021/ja105296a] [PMID: 20853844]

[79] Umut, E. Surface modification of nanoparticles used in biomedical applications. *Modern surface engineering treatments.,* **2013**, *20*, 185-208.
[http://dx.doi.org/10.5772/55746]

[80] Albayrak Ari, G.; Aydin, I. Nanocomposites prepared by solution blending: microstructure and mechanical properties. *J. Macromol. Sci. Part B Phys.,* **2008**, *47*(2), 260-267.
[http://dx.doi.org/10.1080/00222340701748743]

[81] Choudhary, V; Gupta, A. Polymer/carbon nanotube nanocomposites. *Carbon nanotubes-polymer nanocomposites,* **2011**, *11*, 65.

[82] Fawaz, J; Mittal, V Synthesis of polymer nanocomposites: review of various techniques. *Synthesis techniques for polymer nanocomposites.,* **2014**, *24*, 1.
[http://dx.doi.org/10.1002/9783527670307.ch1]

[83] Maiti, S.; Suin, S.; Shrivastava, N.K.; Khatua, B.B. Low percolation threshold in melt-blended PC/MWCNT nanocomposites in the presence of styrene acrylonitrile (SAN) copolymer: Preparation and characterizations. *Synth. Met.,* **2013**, *165*, 40-50.
[http://dx.doi.org/10.1016/j.synthmet.2013.01.007]

[84] Cai, Y.; Jiang, J.S.; Zheng, B.; Xie, M.R. Synthesis and properties of magnetic sensitive shape memory Fe $_3$ O $_4$ /poly(ε-caprolactone)-polyurethane nanocomposites. *J. Appl. Polym. Sci.,* **2013**, *127*(1), 49-56.
[http://dx.doi.org/10.1002/app.36849]

[85] Deng, Y.; Li, Y.; Dai, J.; Lang, M.; Huang, X. Functionalization of graphene oxide towards thermo-sensitive nanocomposites *via* moderate *in situ* SET-LRP. *J. Polym. Sci. A Polym. Chem.,* **2011**, *49*(22),

4747-4755.
[http://dx.doi.org/10.1002/pola.24919]

[86] Luan, J.; Wang, S.; Hu, Z.; Zhang, L. Synthesis techniques, properties and applications of polymer nanocomposites. *Curr. Org. Synth.,* **2012**, *9*(1), 114-136.
[http://dx.doi.org/10.2174/157017912798889161]

[87] Malin, F.; Znoj, B.; Šegedin, U.; Skale, S.; Golob, J.; Venturini, P. Polyacryl–nanoclay composite for anticorrosion application. *Prog. Org. Coat.,* **2013**, *76*(10), 1471-1476.
[http://dx.doi.org/10.1016/j.porgcoat.2013.06.004]

[88] Yang, Y.; Romano, M.; Feng, G.; Wang, X.; Wu, T.; Holdren, S.; Zachariah, M.R. Growth of sub-5 nm metal nanoclusters in polymer melt aerosol droplets. *Langmuir,* **2018**, *34*(2), 585-594.
[http://dx.doi.org/10.1021/acs.langmuir.7b02900] [PMID: 29249148]

[89] Wang, L. Differential structure for temperature sensing based on conductive polymer composites. *IEEE Trans. Electron Dev.,* **2015**, *62*(9), 3025-3028.
[http://dx.doi.org/10.1109/TED.2015.2438075]

[90] Matlou, G.G.; Nkosi, D.; Pillay, K.; Arotiba, O. Electrochemical detection of Hg(II) in water using self-assembled single walled carbon nanotube-poly(m -amino benzene sulfonic acid) on gold electrode. *Sens. Biosensing Res.,* **2016**, *10*, 27-33.
[http://dx.doi.org/10.1016/j.sbsr.2016.08.003]

[91] Li, Y.; Ban, H.; Yang, M. Highly sensitive NH3 gas sensors based on novel polypyrrole-coated SnO2 nanosheet nanocomposites. *Sens. Actuators B Chem.,* **2016**, *224*, 449-457.
[http://dx.doi.org/10.1016/j.snb.2015.10.078]

[92] Costa, P.; Carvalho, M.F.; Correia, V.; Viana, J.C.; Lanceros-Mendez, S. Polymer nanocomposite-based strain sensors with tailored processability and improved device integration. *ACS Appl. Nano Mater.,* **2018**, *1*(6), 3015-3025.
[http://dx.doi.org/10.1021/acsanm.8b00647]

[93] Qi, X.; Yao, X.; Deng, S.; Zhou, T.; Fu, Q. Water-induced shape memory effect of graphene oxide reinforced polyvinyl alcohol nanocomposites. *J. Mater. Chem. A Mater. Energy Sustain.,* **2014**, *2*(7), 2240-2249.
[http://dx.doi.org/10.1039/C3TA14340F]

[94] Shi, Q.; Xia, H.; Li, P.; Wang, Y.S.; Wang, L.; Li, S.X.; Wang, G.; Lv, C.; Niu, L.G.; Sun, H.B. Photothermal surface plasmon resonance and interband transition-enhanced nanocomposite hydrogel actuators with hand-like dynamic manipulation. *Adv. Opt. Mater.,* **2017**, *5*(22), 1700442.
[http://dx.doi.org/10.1002/adom.201700442]

[95] Asgari, M.; Soleymani, M.; Miri, T.; Barati, A. Design of thermosensitive polymer-coated magnetic mesoporous silica nanocomposites with a core-shell-shell structure as a magnetic/temperature dual-responsive drug delivery vehicle. *Polym. Adv. Technol.,* **2021**, *32*(10), 4101-4109.
[http://dx.doi.org/10.1002/pat.5417]

[96] Kulkarni, R.V.; Sa, B. Electroresponsive polyacrylamide-grafted-xanthan hydrogels for drug delivery. *J. Bioact. Compat. Polym.,* **2009**, *24*(4), 368-384.
[http://dx.doi.org/10.1177/0883911509104475]

[97] Nobuhiko, Y.; Jun, N.; Teruo, O.; Yasuhisa, S. Regulated release of drug microspheres from inflammation responsive degradable matrices of crosslinked hyaluronic acid. *J. Control. Release,* **1993**, *25*(1-2), 133-143.
[http://dx.doi.org/10.1016/0168-3659(93)90102-B]

[98] Teoh, S.H.; Chia, H.Y.; Lee, M.S.; Nasyitah, A.J.N.; Luqman, H.B.S.M.; Nurhidayah, S.; Tan, W.C.K. Self healing composite for aircraft's structural application. *Int. J. Mod. Phys. B,* **2010**, *24*(01n02), 157-163.
[http://dx.doi.org/10.1142/S0217979210064083]

[99] Oh, Y.; Armstrong, D.L.; Finnerty, C.; Zheng, S.; Hu, M.; Torrents, A.; Mi, B. Understanding the pH-responsive behavior of graphene oxide membrane in removing ions and organic micropollulants. *J. Membr. Sci.,* **2017**, *541*, 235-243. [http://dx.doi.org/10.1016/j.memsci.2017.07.005]

[100] Himstedt, H.H.; Yang, Q.; Qian, X.; Ranil Wickramasinghe, S.; Ulbricht, M. Toward remote-controlled valve functions *via* magnetically responsive capillary pore membranes. *J. Membr. Sci.,* **2012**, *423-424*, 257-266. [http://dx.doi.org/10.1016/j.memsci.2012.08.015]

CHAPTER 2

Shape-Memory Polymer Nanocomposites and their Applications

Pronob Gogoi[1,*] and **Prithiviraj Khakhlary**[2]

[1] *Department of Chemistry, Diphu Government College, Diphu, Karbi Anglong, Assam, India*

[2] *Department of Chemistry, Dibrugarh University, Dibrugarh, Assam, India*

Abstract: Shape-memory polymer nanocomposites (SMPCs) are considered an advanced type of smart material, in which a specific stimulus influences the structure and thus memorizes their unique shape. SMPCs are widely used in high-performance water–vapor permeability materials, sensors and actuators, intelligent medical devices, morphing applications, and self-deployable structures in spacecraft. Modification of SMPCs overcomes the limitations of the pristine polymer and classical composites and can be counted among the main groups of smart materials. The performance of SMPCs under thermal, electric, optical, magnetic, and solvent stimuli can be improved further, potentially giving them a new range of uses. The most extensively researched SMPCs are those based on polyurethane, epoxy, polycaprolactam, polylactic acid, and polyvinyl alcohol; these also include those nanofillers based on carbon like CNTs, carbon black, graphene oxide, graphene nanoplatelets, graphene quantum dots, *etc*; metal oxides like Fe_3O_4 and TiO_2; cellulose like cellulose nanocrystals and nano-cellulose gel; and other nanomaterials such as nano-clay, TiN, AuNRs, organic nanoparticles, silica, sepiolite, silsesquioxane, and hydroxyapatite nanofillers.

Keywords: Advanced materials, Actuators, Fillers, High performance, Intelligent medical device, Nanoparticles, Nanocomposites, Sensors, Stimuli-responsive, Shape memory effect, Shape memory polymer, Smart materials, Self-deployable, Thermo-responsive.

INTRODUCTION

Shape-memory polymers (SMPs) are considered a kind of smart material with tunable size, shape, stiffness, and strain in response to different external stimuli (like heat, electric and magnetic field, water, or light) including physiologic ones such as pH, body temperature and ions concentration. The unique property of shape memory polymers is that they have the capability of storing and recovering

* **Corresponding author Pronob Gogoi:** Department of Chemistry, Diphu Government College, Diphu, Karbi Anglong, Assam, India; Tel: +917002060284; E-mail: gogoipronob@gmail.com

Partha Pratim Saikia & Bhaskar Jyoti Saikia (Eds.)

a large deformation. Furthermore, SMPs are materials with the ability to sense external stimuli and react in a predictable and regulated way. They can be designed to memorize one or more temporary shapes in addition to a permanent shape [1, 2].

A polymer needs both the switch segment and the net point in order to display the shape memory property. The initial shape can be created by the crystalline phase, by cross-linking *via* a covalent connection, or by physical means and an interpenetrated polymer network that is determined by the net point of SMPs. The net point must remain constant through the formation of a stable polymer structure in order to maintain the material's shape even at high temperatures. The switch segment, in particular the flexible portions of the polymer matrix for reversible switching transition, can be a physical transition and can restore the temporary structure of SMPs. Moreover, a physical phenomenon like crystallization or glass transition of the polymers and chemical reactions like oxidation or reduction, Diels–Alder reaction, supramolecular association or disassociation, self-assembly of metal-ligand coordination, *etc.*, influence the shape memory properties of the SMPs [3, 4]. Based on the development of molecular mechanisms, the comprehensive architecture of the SMPs is presented in Fig. (**1**) [5].

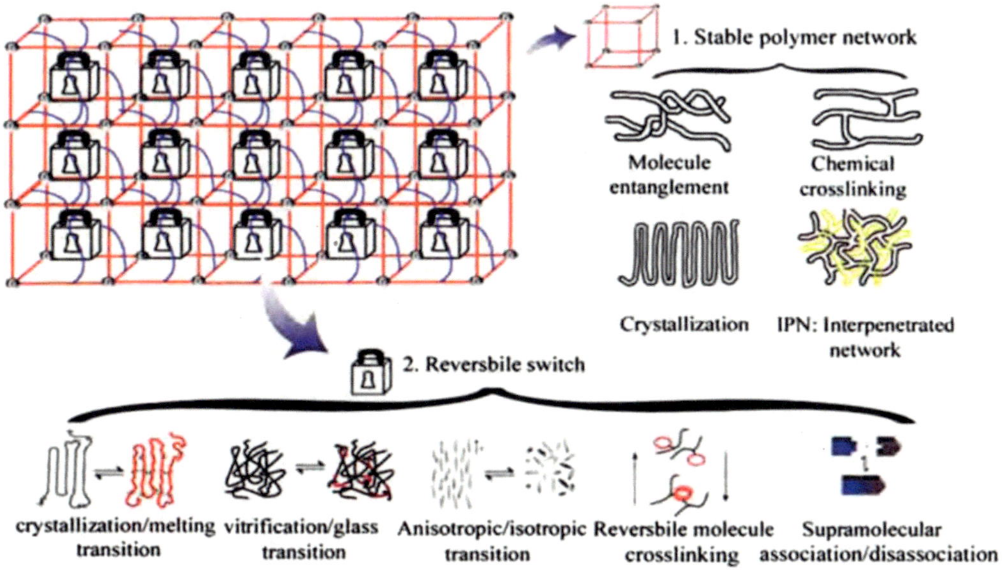

Fig. (1). Various molecular structures of SPMs [5].

SMPs are frequently used in high-performance water-vapor permeability materials, self-deployable structures in airplanes, morphing applications, textiles and clothing, sensors and actuators, and smart sophisticated medical equipment. However, the high-performance applications of SPMs are limited by low heat conductivity, low recovery stress, low tensile strength and stiffness, and inertness to electrical and light stimuli [3, 4]. Among all the SMPs, epoxy, polyurethane, polylactic acid, polycaprolactam, and polyvinyl alcohol-based SMPs are widely studied.

Due to nanofillers' strong polymer interaction and wide surface area, researchers have focused more on shape memory polymer nanocomposites (SMPCs) recently than on traditional composites. This way of preparation of SMPCs is an effort to address the shortcomings of SMPs [6]. Nanofillers such as metal oxide, nanotubes, nanofibers, nanospheres, nanorods, and so on can be incorporated into a polymer matrix to create a variety of SMPCs with specific properties. The decoration of SMPCs can enhance the performance characteristics of the SMPCs under electric, light, magnetic, thermal, and solvent stimuli, which may open up a new dimension for the SMPCs to be used in more advanced applications [5, 7].

Meticulous designing of SMPCs can improve the performance characteristics of the SPMs and increase the adaptability of stimuli-responsive properties. By adding a modest amount of cross-linking micro- or nano-sized fillers, the mechanical strength and recovery stress of SMPs can be significantly increased. Hence, as compared to the tidy SPMs, the designed SMPCs show noticeably better mechanical and shape memory qualities along with the new properties. Decoration of SMPCs can improve thermal stimuli-active and shape memory properties along with the reinforcement effect of the fillers. Typical examples of unique shape memory effects include stimuli-memory effect, two-way shape memory, multiple-shape memory, spatially controlled shape memory, and self-healing [7].

The fillers in SMPCs can improve the way SPMCs perform by activating the polymer matrix with different energy inputs and giving the composite system varied functions. But there are still certain obstacles to overcome. Due to the substantial or cyclic deformation that SMPCs endure, the primary problem in the manufacture of SMPCs is achieving homogeneous mixing of the fillers and SMP matrix. The mechanical characteristics of SMPCs are diminished by the modest intermolecular forces of attraction between the fillers and the matrix. Additionally, this may result in the Mullins effect of filled composite materials, which lowers the mechanical qualities of the polymer composites that are the outcome of the first extension. The overall performance characteristics of the SMPCs are influenced by the filler's geometries, modulus, volume contents, surface qualities, and their distributions; these features can be used to design and optimize the performance parameters of decorated SMPCs [5 - 7].

The designing of SMPCs results in the reduction of shape recovery time due to the increase in thermal conductivity, increase in mechanical properties, and recovery stress due to the reinforcing effect of the fillers, easy tuning of switching temperature, mechanical properties, and biomedical properties as compared to the pristine SMPs. Moreover, the SMPCs became sensitive to electricity, magnetic fields, light, and moisture. That is, SMPCs of desired properties can be prepared by carefully selecting the SMPs and the filler [8 - 10]. Overall, the formulation of SMPCs can enhance the performance characteristics of SMPCs under stimuli such as solvent, light, magnetic, electric, and thermal which may open up a new aspect of SMPCs to be used in more advanced applications as shown in Fig. (**2**).

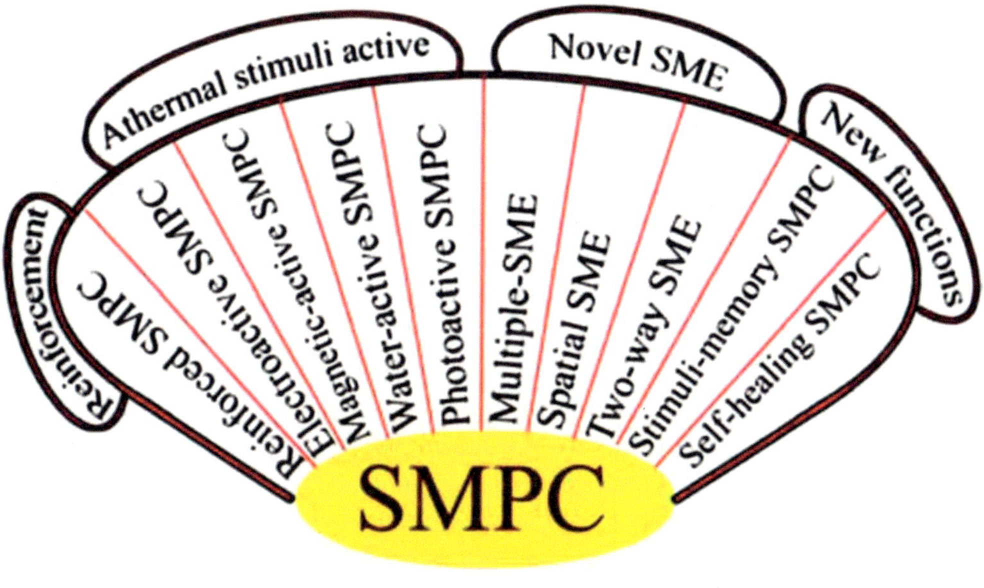

Fig. (2). SMPCs with a thermal stimuli-active effect, SMPCs with novel SME, and SMPCs with new functions [5].

So far, the preparation of a wide range of SMPCs has been reported and details of the performance characteristics have been studied. It is observed that depending on the nature of SPMs and fillers used, they possess different properties and are suitable for specific filed applications. Moreover, depending on the applications of the SMPCs, the SMPCs of desired properties can be prepared. In this regard, SMPCs of Polyurethane (PU), Epoxy, Poly(ε-caprolactone) (PCL), Poly(lactic acid) (PLA), Poly(vinyl alcohol) (PVA), and carbon-based nanoparticles like GO,

CNTs, CDs, carbon black, *etc*, metal oxide-based nanoparticles like TiO_2, Fe_3O_4, Al_2O_3, *etc*, cellulose-based nanoparticles like cellulose nanocrystals and nano-cellulose gel, and other nanomaterials like nano-clay, silica, TiN, AuNR, silsesquioxane, hydroxyapatite nanofillers, and some organic nanoparticles are broadly discussed [11 - 16].

SMPCs with Carbon-Based Nanofiller

A wide range of SMPCs with carbon-based nanofillers like CNTs, carbon black, graphene oxide, and graphene with SMPs like polyurethane (PU), epoxy, poly(lactic acid) poly(vinyl alcohol) (PVA) are reported [12, 14]. One of the main objectives of incorporating CNTs or functionalized CNTs into the SMP matrix is to achieve an electroactive shape memory effect by Joule heating in intrinsic thermal-responsive SMPs. The biggest challenge of SMP/CNT nanocomposites is to disperse the CNTs uniformly in the SMP matrix. Many methods have been used such as melt mixing followed by *in-situ* polymerization, chaotic mixing followed by *in-situ* polymerization, solution mixing assisted with ultrasonic distribution, and chemical functionalization of CNTs followed by cross-linking reaction. The PVA/CNTs SMPCs that were made utilizing a unique coagulation spinning technology have the highest recovery stress. A very high loading of CNTs into the PVA matrix was made possible by the technique where the CNTs are stabilized by the application of a surfactant and then injected into the co-flowing stream of the coagulating PVA solution [5, 17].

SMPCs with Exfoliated Nanoclay

Exfoliated nanoclay-based SMPCs target the modification of nanoclays to facilitate their reaction with the shape memory polymer matrix. In other words, the included nano clay is essential to increasing the shape memory capabilities of the composites. Chemically modified organic-exfoliated nano clay can significantly improve the shape recovery ratios, shape recovery stress, and mechanical properties of the SMPCs [18]. Functionalization of the nano clay not only facilitates the homogeneous mixing with the SPM matrix but also adds an additional network contributing to the shape memory effect of the SMPCs. The heat treatment's loss of moisture and the majority of surface hydroxyl groups can increase the interfacial interaction between the SMP and clay contact, making the clay more compatible with the SMP matrix. These shape memory effects and mechanical properties are reported to be significantly improved in such SMPCs [19, 20].

SMPCs with SiC or SiO_2 Particles

SiC and SiO_2 widely available in nature, show spheres and rod geometries. Their presence dramatically improves the mechanical characteristics of SMPCs while lowering reaction times. A study showed that with the clay contents in the SMPCs, the free recovery speed, micro-hardness, and elastic modulus of the prepared nanocomposites increase. However, the mentioned relationship follows up to a certain level of clay loading after that, the nanocomposites show irrecoverable deformation. However, the excellent interface between fillers and matrix is required to achieve SMPCs with high shape recovery and shape fixity ratio [21, 22].

SMPCs with Metals and Metal Oxides

Metals and their oxides usually possess good electrical, magnetic, thermal, and ductility properties and are hence chosen to fabricate flexible electronic devices. They provide choices to fabricate SMPCs with multi-stimulus methods, which can be used to trigger a multi-shape recovery process. The SMPs are capable of memorizing multi-shape memory effects where different shapes can be regained by changing the temperature range [23]. The reported multi-shape memory effects of the SMPCs are caused by changes in distinct relaxation modes, either inactive or active, in response to temperature changes. With the manipulation of magnetic strength at varying temperatures within the SMPCs, further diverse forms can be created through magnetic-thermal conversion [23].

Applications

Thorough research on shape memory polymers and their nanocomposites has proven to have a wide range of practical field applications in smart textiles, soft robotics, aircraft, flexible electronics, advanced medical devices, controlled drug delivery, shape-memory arrays, self-deployable structures, sensors, and actuators, high-performance materials with water vapor permeability, 3D printing, *etc* [20]. Using materials that melt or cure photonically, 3D printing, also known as additive manufacturing, may quickly create three-dimensional objects layer by layer from computer-aided designs. It is possible to perform very sophisticated shape modifications and self-driving processes by integrating 3D printing technology with SMPCs. A multi-material 3D printer with customizable material parameters is depicted in Fig. (**3**) [23].

SMPCs are good candidates for specific applications in spacecraft, mainly due to their low density, large recoverable strain, and long recovery time, which results in a small impact on spacecraft release mechanisms. SMPCs are able to produce active deformation, which are essential characteristics for use in aerospace engineering, especially in deployable structures. Typical examples include hinges, gravity gradient booms, deployable panels, and reflector antennas. SMPC hinges reinforced with carbon fiber plain-weave fabrics have been developed to overcome the problems associated with traditional metallic hinges that take up a lot of space and weight. The hinge displayed a shape recovery ratio of about 100% and could actuate a prototype solar array during the shape recovery process [24].

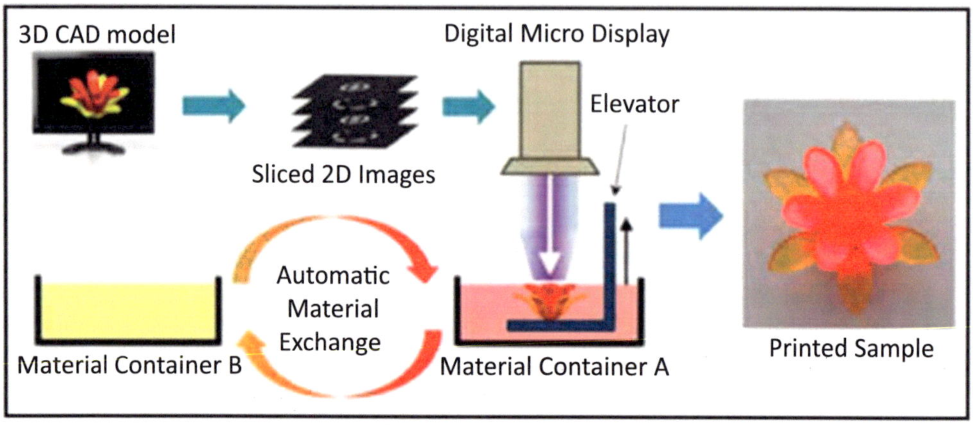

Fig. (3). A workflow illustrates the process of fabricating a multi-material structure [23].

SMPCs have advanced significantly over the last ten years, and they may find use in the fields of bionics, energy, and civil engineering. SMPCs open up a wide range of new SMPC possibilities, including multiple-shape memory polymer (MSMP) composites, two-way SMPCs, spatially controlled SMPCs, electroactive SMPCs, photoactive SMPCs, stimuli-memory effect, and self-healing SMPCs [5]. With the advancement of composite design and manufacturing technology, the distinctions between functional materials and structural materials are blurred. The SMPs and SMPCs can be used to change the deformation of structures, during which the electronic devices keep working. Shape memory structures and integrated circuits can be used to provide an endless number of possible uses for SMPCs. One of the most important uses of SMPCs is in the development of self-folding origami energy harvesting systems [23].

High strength and elastic modulus fiber-reinforced SMPCs can greatly boost their mechanical strength; as a result, they are utilized in spaceship vibration control and self-deployable structures [24 - 28]. The practical potential of SMPCs has been strongly linked to their advancement. For minimally invasive surgery, they might be utilized as a self-tightening suture [29]. The SMPC suture was first stretched into an elongated shape, and it was then placed loosely to the wound while in this temporary shape. The heat from the human body then caused the suture to regain its shape, which caused the suture to shrink and the knot to tighten on its own without the need for outside assistance. Utilizing the material's polymeric nature, further capabilities are added, like biodegradability, which does away with the need for a separate suture removal procedure. Significant interest in additional biomedical applications of SMPCs has been sparked by this prototype demonstration. A large share of SMP research in recent years has focused on customizing SMPCs for diverse biomedical applications. The SMPCs find increased biomedical applications like surgical and vascular for the replacement of hand-held surgery instruments due to their active deformation ability [24]. The fact that SMP and SMPCs have a high value that is unmatched by the majority of non-biomedical applications is one reason for the intensive focus on these applications in the biomedical field [16, 29].

SMPCs resemble biomaterials and they can be thought of as shape memory bio-inspired materials and metamaterials like honeycomb, spider silk, human teeth, and bone. The metamaterials exhibit deformation, geometries, and properties of the repeating units and can be applied to attain auxetics, wave guidance, and programmable mechanical properties. Multilateral 3D printing technology with thermally tunable auxetic behaviours can be achieved by joining structural design with metamaterials [30 - 32].

Flexible electronics based on SMP and SMPCs have become a very interesting area of research. As these materials are flexible and have enough load-bearing capacity hence, they can be used as the substrate to develop soft electronic devices. The polymer matrix of SMPCs becomes softened at a glass transition temperature (Tg) to adjust to the rigid electronics and can maintain stable properties after multiple bending-recovery tests that are essential for wearable devices [24].

By virtue of the properties exhibited by the SMPCs, they find suitable applications in soft robotics. Conventional robots are usually heavy, rigid, bulk and not capable of interactions with humans. SMPCs are able to overcome those serious problems associated with conventional materials. A wide variety of materials have been developed, promoting the improvement of soft robotics and bringing new possibilities in solving problems that could, otherwise, not be possible by conventional robots. The lightweight, highly flexible, and mechanically compliant complex elastomers are used to make soft robots. In addition, inflectable SMPCs are used to develop soft crawling robots, actuators, grippers and other complex robots that have great potential to complete complex tasks like drug delivery, search and rescue, and acting as human assistants [24, 33].

CONCLUDING REMARKS

Much work has been done on developing the intricate and regulated shape recovery effect of SMPs and SMPCs during the past few decades. Selective heating is one such intricate and spatially controlled form of recovery method. The incorporation of a small number of fillers can drastically enhance the mechanical strength and recovery stress of SMPCs along with the introduction of new properties and hence broaden their applications. SMPCs exhibit improved athermal stimuli-active effects, shape memory effects, and mechanical and load-bearing properties. With these, SMPCs have been considered advanced types of smart materials, in which a specific stimulus influences the structure and thus memorizes their unique shape. SMPCs are widely used in high-performance water-vapour permeability materials, sensors and actuators, intelligent medical devices, morphing applications, and self-deployable structures in spacecraft. However, understanding the structure-property relationship of the SMPCs will help in developing and designing potentially more advanced field applications like spatially controlled SMPCs, electroactive SMPCs, photoactive SMPCs, stimuli-memory effect, self-healing SMPCs, *etc*.

Research will continue to focus on developing SMPCs with features that are tailored to the unique needs of particular applications. The findings of new SMPCs will be important since they will expand the material's capabilities and may have a more profound effect on how they are used. The overall properties of SMPCs are influenced by the filler's shape and size, surface properties, and their distributions in the polymer matrix. The development of patterned SMPCs with better electrical and mechanical properties is one of the biggest achievements over the last decade of advanced manufacturing technologies that can provide a specific function of self-folding from a two-dimensional sheet to a three-dimensional structure.

REFERENCES

[1] Tao, X. *Smart fibres, fabrics and clothing*; Woodhead Publishing Limite: Cambridge, **2001**. [http://dx.doi.org/10.1533/9781855737600]

[2] Sun, L.; Huang, W.M.; Ding, Z.; Zhao, Y.; Wang, C.C.; Purnawali, H.; Tang, C. Stimulus-responsive shape memory materials: A review. *Mater. Des.*, **2012**, *33*, 577-640. [http://dx.doi.org/10.1016/j.matdes.2011.04.065]

[3] Bhanushali, H.; Amrutkar, S.; Mestry, S.; Mhaske, S.T. Shape memory polymer nanocomposite: a review on structure–property relationship. *Polym. Bull.*, **2022**, *79*(6), 3437-3493. [http://dx.doi.org/10.1007/s00289-021-03686-x]

[4] Wang, W.; Liu, Y.; Leng, J. Recent developments in shape memory polymer nanocomposites: Actuation methods and mechanisms. *Coord. Chem. Rev.*, **2016**, *320-321*, 38-52. [http://dx.doi.org/10.1016/j.ccr.2016.03.007]

[5] Meng, H.; Li, G. A review of stimuli-responsive shape memory polymer composites. *Polymer (Guildf.)*, **2013**, *54*(9), 2199-2221.

[http://dx.doi.org/10.1016/j.polymer.2013.02.023]

[6] Liu, T.; Zhou, T.; Yao, Y.; Zhang, F.; Liu, L.; Liu, Y.; Leng, J. Stimulus methods of multi-functional shape memory polymer nanocomposites: A review. *Compos., Part A Appl. Sci. Manuf.,* **2017**, *100*, 20-30.
[http://dx.doi.org/10.1016/j.compositesa.2017.04.022]

[7] Leng, J.; Lan, X.; Liu, Y.; Du, S. Shape-memory polymers and their composites: Stimulus methods and applications. *Prog. Mater. Sci.,* **2011**, *56*(7), 1077-1135.
[http://dx.doi.org/10.1016/j.pmatsci.2011.03.001]

[8] Hornbogen, E. Comparison of shape memory metals and polymers. *Adv. Eng. Mater.,* **2006**, *8*(1-2), 101-106.
[http://dx.doi.org/10.1002/adem.200500193]

[9] Kausar, A. Shape memory interpenetrating network hybrids of epoxy/poly(urea-amide) and organic nanoparticle. *Journal of the Chinese Advanced Materials Society,* **2017**, *5*(3), 158-173.
[http://dx.doi.org/10.1080/22243682.2017.1329028]

[10] Tian, G.; Zhu, G.; Xu, S.; Ren, T. A novel shape memory poly(ε-caprolactone)/hydroxyapatite nanoparticle networks for potential biomedical applications. *J. Solid State Chem.,* **2019**, *272*, 78-86.
[http://dx.doi.org/10.1016/j.jssc.2019.01.029]

[11] Navarro-Baena, I.; Arrieta, M.P.; Sonseca, A.; Torre, L.; López, D.; Giménez, E.; Kenny, J.M.; Peponi, L. Biodegradable nanocomposites based on poly(ester-urethane) and nanosized hydroxyapatite: Plastificant and reinforcement effects. *Polym. Degrad. Stabil.,* **2015**, *121*, 171-179.
[http://dx.doi.org/10.1016/j.polymdegradstab.2015.09.002]

[12] Burgos, N.; Tolaguera, D.; Fiori, S.; Jiménez, A. Synthesis and characterization of lactic acid oligomers: evaluation of performance as poly(lactic acid) plasticizers. *J. Polym. Environ.,* **2014**, *22*(2), 227-235.
[http://dx.doi.org/10.1007/s10924-013-0628-5]

[13] Arrieta, M.P.; Fortunati, E.; Dominici, F.; López, J.; Kenny, J.M. Bionanocomposite films based on plasticized PLA–PHB/cellulose nanocrystal blends. *Carbohydr. Polym.,* **2015**, *121*, 265-275.
[http://dx.doi.org/10.1016/j.carbpol.2014.12.056] [PMID: 25659698]

[14] Ji, M.; Deng, H.; Yan, D.; Li, X.; Duan, L.; Fu, Q. Selective localization of multi-walled carbon nanotubes in thermoplastic elastomer blends: An effective method for tunable resistivity–strain sensing behavior. *Compos. Sci. Technol.,* **2014**, *92*, 16-26.
[http://dx.doi.org/10.1016/j.compscitech.2013.11.018]

[15] Teymouri, M.; Kokabi, M.; Alamdarnejad, G. Conductive shape-memory polyurethane/multiwall carbon nanotube nanocomposite aerogels. *J. Appl. Polym. Sci.,* **2019**, *48602*, 1-11.

[16] Xie, T. Recent advances in polymer shape memory. *Polymer (Guildf.),* **2011**, *52*(22), 4985-5000.
[http://dx.doi.org/10.1016/j.polymer.2011.08.003]

[17] Miaudet, P.; Derré, A.; Maugey, M.; Zakri, C.; Piccione, P.M.; Inoubli, R.; Poulin, P. Shape and temperature memory of nanocomposites with broadened glass transition. *Science,* **2007**, *318*(5854), 1294-1296.
[http://dx.doi.org/10.1126/science.1145593] [PMID: 18033882]

[18] Kim, M.S.; Jun, J.K.; Jeong, H.M. Shape memory and physical properties of poly(ethyl methacrylate)/Na-MMT nanocomposites prepared by macroazoinitiator intercalated in Na-MMT. *Compos. Sci. Technol.,* **2008**, *68*(7-8), 1919-1926.
[http://dx.doi.org/10.1016/j.compscitech.2007.12.015]

[19] Xu, B.; Huang, W.M.; Pei, Y.T.; Chen, Z.G.; Kraft, A.; Reuben, R.; De Hosson, J.T.M.; Fu, Y.Q. Mechanical properties of attapulgite clay reinforced polyurethane shape-memory nanocomposites. *Eur. Polym. J.,* **2009**, *45*(7), 1904-1911.
[http://dx.doi.org/10.1016/j.eurpolymj.2009.04.004]

[20] Xiao, Y.; Zhou, S.; Wang, L.; Zheng, X.; Gong, T. Crosslinked poly(ε-caprolactone)/poly(sebacic anhydride) composites combining biodegradation, controlled drug release and shape memory effect. *Compos., Part B Eng.,* **2010**, *41*(7), 537-542. [http://dx.doi.org/10.1016/j.compositesb.2010.07.001]

[21] Boudjellal, A.; Trache, D.; Khimeche, K.; Hafsaoui, S.L.; Bougamra, A.; Tcharkhtchi, A.; Durastanti, J.F. Stimulation and reinforcement of shape-memory polymers and their composites: A review. *Journal of Thermoplastic Composite Materials,* **2022**, *35*(11), 2227-2260. [http://dx.doi.org/10.1177/0892705720930775]

[22] Wang, Y.; Tian, W.; Liu, X.; Ye, J. Thermal sensitive shape memory behavior of epoxy composites reinforced with silicon carbide whiskers. *Appl. Sci. (Basel),* **2017**, *7*(1), 108. [http://dx.doi.org/10.3390/app7010108]

[23] Lei, M.; Chen, Z.; Lu, H.; Yu, K. Recent progress in shape memory polymer composites: methods, properties, applications and prospects. *Nanotechnol. Rev.,* **2019**, *8*(1), 327-351. [http://dx.doi.org/10.1515/ntrev-2019-0031]

[24] Xia, Y.; He, Y.; Zhang, F.; Liu, Y.; Leng, J. A review of shape memory polymers and composites: Mechanisms, materials, and applications. *Adv. Mater.,* **2021**, *33*(6), 2000713. [http://dx.doi.org/10.1002/adma.202000713] [PMID: 32969090]

[25] Lan, X.; Liu, Y.; Lv, H.; Wang, X.; Leng, J.; Du, S. Fiber reinforced shape-memory polymer composite and its application in a deployable hinge. *Smart Mater. Struct.,* **2009**, *18*(2), 024002. [http://dx.doi.org/10.1088/0964-1726/18/2/024002]

[26] Jung, Y.C.; Sahoo, N.G.; Cho, J.W. Polyurethane-carbon nanotube nanocomposites prepared by *in-situ* polymerization with electroactive shape memory. *Macromol. Rapid Commun.,* **2006**, *27*, 126-131. [http://dx.doi.org/10.1002/marc.200500658]

[27] Meng, Q.; Hu, J. A review of shape memory polymer composites and blends. *Compos., Part A Appl. Sci. Manuf.,* **2009**, *40*(11), 1661-1672. [http://dx.doi.org/10.1016/j.compositesa.2009.08.011]

[28] Ohki, T.; Ni, Q.Q.; Ohsako, N.; Iwamoto, M. Mechanical and shape memory behavior of composites with shape memory polymer. *Compos., Part A Appl. Sci. Manuf.,* **2004**, *35*(9), 1065-1073. [http://dx.doi.org/10.1016/j.compositesa.2004.03.001]

[29] Hardy, J.G.; Palma, M.; Wind, S.J.; Biggs, M.J. Responsive biomaterials: Advances in materials based on shape-memory polymers. *Adv. Mater.,* **2016**, *28*(27), 5717-5724. [http://dx.doi.org/10.1002/adma.201505417] [PMID: 27120512]

[30] Baer, G.M.; Wilson, T.S.; Small, W., IV; Hartman, J.; Benett, W.J.; Matthews, D.L.; Maitland, D.J. Thermomechanical properties, collapse pressure, and expansion of shape memory polymer neurovascular stent prototypes. *J. Biomed. Mater. Res. B Appl. Biomater.,* **2009**, *90B*(1), 421-429. [http://dx.doi.org/10.1002/jbm.b.31301] [PMID: 19107804]

[31] Zhao, Q.; Qi, H.J.; Xie, T. Recent progress in shape memory polymer: New behavior, enabling materials, and mechanistic understanding. *Prog. Polym. Sci.,* **2015**, *49-50*, 79-120. [http://dx.doi.org/10.1016/j.progpolymsci.2015.04.001]

[32] Zhang, Y.; Zhang, F.; Yan, Z.; Ma, Q.; Li, X.; Huang, Y.; Rogers, J.A. Printing, folding and assembly methods for forming 3D mesostructures in advanced materials. *Nat. Rev. Mater.,* **2017**, *2*(4), 17019. [http://dx.doi.org/10.1038/natrevmats.2017.19]

[33] Xiong, J.; Chen, J.; Lee, P.S. Functional fibers and fabrics for soft robotics, wearables, and human–robot interface. *Adv. Mater.,* **2021**, *33*(19), 2002640. [http://dx.doi.org/10.1002/adma.202002640] [PMID: 33025662]

CHAPTER 3

Smart Polymeric Composites in Controlled Drug Delivery

Kiranjyoti Mohan[1,*]

[1] *Department of Chemistry, Rangapara College, Sonitpur-784505, Assam, India*

Abstract: A drug delivery system (DDS) is the mode by which a drug is incorporated inside the body to produce therapeutic effects at the desired pathological site. Common DDSs include capsules and tablets (through the oral route), intravenous injections, nasal and ocular drops, transdermal patches, ointments, gels, *etc*. The important criteria for an active pharmaceutical ingredient (API) to be effective are that it should be precise and specific for the site of interest, with minimal side effects. This is where the formulation of the drug becomes very important. Regulating the entry of the drug into the body, followed by excellent distribution is necessary to elucidate a quick reaction. Hence, substantial consideration has to be given to the absorption, distribution, metabolism, and excretion (ADME) of the drug. Polymers have been integrated with APIs mostly as inert drug carriers to facilitate ADME, with the pre-requisites that the incorporated polymers must be biocompatible, have the ability to improve the drug stability and release kinetics, and avoid accumulation in the bloodstream. To further improve their efficacy, smart polymers have also come into play. They react to environmental stimuli like temperature, light, magnetic and electric fields, *etc*. Polymeric magnetic nanocomposites have shown great promise due to their quick response to magnetic cues and remote-control capabilities. The functionalization of both natural and synthetic biodegradable polymers with superparamagnetic iron oxide nanoparticles (SPIONs) results in DDSs where drugs can be remotely released in a controlled manner.

Keywords: Chemically controlled, Core-shell structure, Diffusion-controlled, Drug delivery system, Magnetic hyperthermia, Magnetically-guided delivery, Multi stimuli-responsive, pH-responsive, Polymer-drug carriers, Smart polymers, Solvent-activated, Superparamagnetic iron oxide nanoparticles, Temperature responsive.

* **Corresponding author Kiranjyoti Mohan:** Department of Chemistry, Rangapara College, Sonitpur-784505, Assam, India; Tel: +91-6002798962; E-mail: mohan.kiran321@gmail.com

Partha Pratim Saikia & Bhaskar Jyoti Saikia (Eds.)

INTRODUCTION

Overview of Drug Delivery Systems

A drug delivery system (DDS) is the method of administering a pharmaceutical formulation inside the body to realize a therapeutic effect. Once a novel drug is developed in the laboratory, the chances of it being solely administered are very rare. Instead, the drug (or more aptly called the active pharmaceutical ingredient (API)) is given as a part of a formulation, combined with inert pharmaceutical ingredients called excipients. The excipients serve very specialized functions and are a part of the drug delivery process. Thus, a DDS can also be described as a means for the drugs to be packaged in such a way that they are protected from physiological degradation and can be delivered at the target site safely at a controlled rate. DDSs can be categorized in two ways: (a) based on the mode of administration of the drug into the body and (b) the intended effect at the target site upon delivery. Both the mode of administration and the intended effect can be local or systemic [1]. Local administration of the DDS is mainly achieved by directly applying it to the target site or introducing it into the circulation system in such a way as to obtain a high concentration of the API at the desired location without exposing other body parts. Local administration routes mainly include applying topical ointments, lotions, creams, and gels transdermally, and also through nasal and eye drops, *etc*. Although advantageous, local administration in most cases is difficult due to accessibility issues arising out of anatomical barriers. In such cases, systemic administration is required where the DDS is applied at an off-target accessible location *via* oral tablets and capsules, intravenous injections, *etc*. The DDS then circulates through the body to reach the intended target site. Due to circulation, the possibility of non-intended organs getting exposed to the drugs increases, leading to higher chances of side effects and toxicity. Once delivered, the drug can produce either localized or systemic effects. Localized effect entails drug activity typically at the site of DDS administration (local effect by local administration). However, it is also possible to obtain local effects from systemic administration by causing the DDS circulating in the bloodstream to accumulate at and affect the target site. If the drug affects the entirety of the body during circulation, this is termed a systemic effect by systemic administration.

During the initial days of therapeutics, APIs were mainly created using small molecules (sizes < 900 Daltons) like steroids and antibiotics. They can effortlessly diffuse through various biological membranes but for that to happen, they must dissolve easily in the biological fluids, which is often not the case. As the therapeutic landscape evolved, other classes of APIs were introduced, including proteins and peptides, antibodies, nucleic acids, live cells, *etc* [2]. These drugs serve new therapeutic functions, but they are also riddled with some difficulties of their own, like problems with stability, delivery conditions, *etc*. In response to the hardships, DDSs have also advanced in order to solve these solubility and stability issues by modifying the local microenvironment and physicochemical properties, and also improve bioavailability, and broadening the activity of the API. An important component of the DDS that is instrumental in solving these issues is the excipient. Additionally, some excipients also act as drug carriers. As the name suggests, a drug carrier acts as a protective barrier for the transportation of pharmaceutical molecules to the target site, without causing any degradation or leakage during the entire process of administration, circulation, and delivery of the drug. An ideal drug carrier should have the ability to control and optimize the rate of drug delivery and increase its duration of activity so that the frequency of drug administration can be reduced.

Smart Polymers as Drug Carriers

Biocompatible polymers have been widely used in DDSs in various ways. They can either be non-biodegradable or biodegradable. In the case of non-biodegradable polymers, surgery is often required to excrete them out of the body once their purpose is completed. Hence, the use of biodegradable polymers is favored. Biodegradable polymers decompose into non-toxic and regular metabolic products through hydrolysis or enzyme-catalyzed reaction. They are classified into two categories: natural polymers, such as cellulose, chitosan, polypeptides and proteins, and synthetic polymers including poly (glycolic-co-lactic) acid (PLGA), poly (lactic acid) (PLA), poly (glycolic acid) (PGA), poly (caprolactone) (PCL), poly (ethylene glycol) (PEG), poly (vinyl pyrrolidone) (PVP), poly (methyl methacrylate) (PMMA), aliphatic polycarbonates and polyphosphazene. Polymers incorporated with drugs can modify the circulation characteristics of the drugs, and degrade at the target site to improve drug release and avoid carrier build-up. Additionally, polymers themselves can sometimes be bioactive and offer therapeutic advantages. Synthetic polymers are preferred over natural ones since it is easy to control the polymer's three-dimensional structure and chemical composition during synthesis. In that way, polymers can be tailor-made to fulfill desired requirements. For instance, polymers can be synthesized with their functional group orientation in a particular direction so that they interact strongly with the API. Synthetic polymers, however, are not always degradable. In such cases, they must be eliminated from the body through renal excretion. So, consideration has to be given to maintaining their molecular weight under the renal excretion threshold.

Polymers can be incorporated with the APIs in a variety of ways: polymers can form covalent bonds with the drug to form a polymer-drug conjugate, or the drug can disperse in the polymer network to form a matrix system, or the drug core can be enclosed inside a polymer coating. Once the drug loading is complete, and the DDS is administered and circulated to the target site, its release from the polymer is also controlled by different mechanisms depending on certain conditions. These include diffusion-controlled, solvent-activated, chemically controlled or externally triggered systems [3].

Diffusion-Controlled DDS

Diffusion-controlled DDSs are normally very simple. The drug is dissolved in a fully swollen polymeric membrane or a matrix that does not degrade during its therapeutic period, *i.e.*, the DDSs do not change their volume, either by swelling or degradation, during the course of the drug release, which is governed by Fick's laws of diffusion. Depending on the type of polymeric structure, diffusion-controlled DDSs are categorized as reservoir type (containing water-insoluble polymeric membranes) and monolithic or matrix type (containing polymeric matrices) systems (Fig. **1**). In reservoir type systems, the drugs, in the form of powder or liquid, are encapsulated inside a semi-permeable polymeric membrane as a reservoir. The drug molecules partition within the polymeric membrane, and then subsequently release into the dissolution medium through the membrane by diffusion. The release rate is determined by the thickness and porosity of the membrane, and also physicochemical properties of the drugs [4]. The membranes are made using a variety of polymers like poly (vinyl acetate) (PVA), polyurethane (PU), hydroxypropyl cellulose (HPC), and ethyl cellulose, for use in the treatment of cancer, birth control, *etc* [5]. For instance, Kim *et al.* successfully demonstrated the permeation of the drugs, riboflavin, and insulin, through a membrane composed of cross-linked PVA/chitosan (CS) blend [6]. Permeation of the drugs in this case was pH dependent, *i.e.*, the drugs showed better permeability in acidic solution, which was attributed to water content as well as water structure (free and frozen bound water) inside the swollen membrane.

In matrix-type systems, the drug is dissolved/ dispersed homogenously in a polymer matrix. Initially, the external layer of the polymer matrix is exposed to the dissolution fluid, and the drug particles present there diffuse out of the DDS. With time, the solution moves to the core, allowing for the drugs present inside to start diffusing outside the matrix [7]. Commonly used polymers as matrix include PVP, guar gum, polyacrylamide (PAAm), *etc*. Khan and Zhu prepared tablets of ibuprofen drug with Carbopol® 934P polymer, and also with a blend of Carbopol® 934P and 971P resins [8]. They observed that a higher amount of Carbopol® 934P in the tablets resulted in the linearization of the drug release curve. On the other hand, the blended mixtures of Carbopol® 934P and 971P resins could produce a more linear drug release curve even at lower polymer levels.

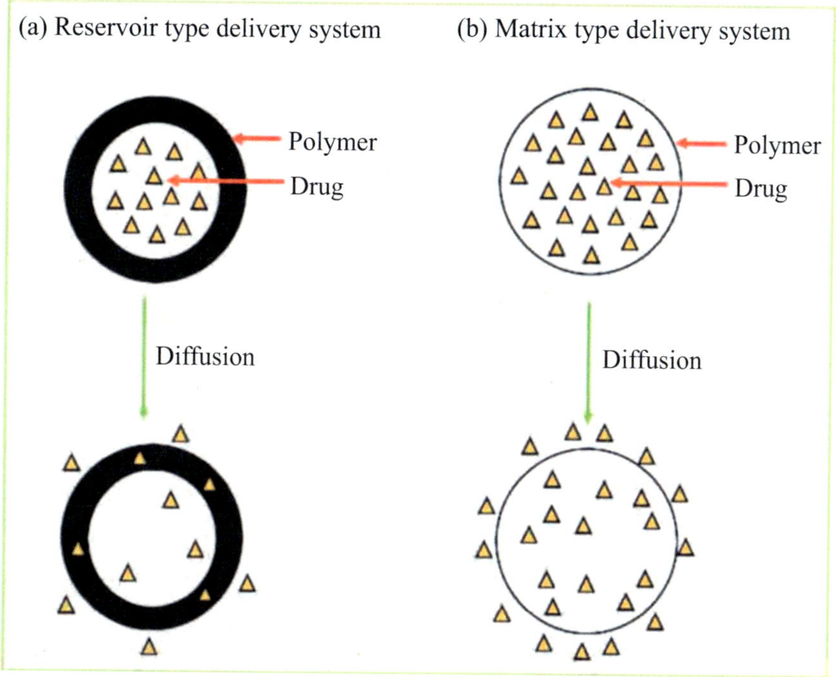

Fig. (1). Diffusion-controlled DDSs: **(a)** reservoir type and **(b)** matrix type delivery systems.

Solvent-Activated DDS

These are divided into swelling-controlled and osmotic pressure-controlled DDSs (Fig. **2**). The rate of drug release is dependent on the penetration of water into the system. In swelling-controlled DDSs, the drug particles are first loaded into hydrophilic polymeric cross-linked chains, which are in a dehydrated state but have the capacity to absorb water without dissolving [9]. At this stage, the system is below its glass transition temperature (T_g), so it exists in a rigid, glassy state with low diffusivity. Once the system comes in contact with an aqueous solution, water infiltrates the polymer matrix lowering the value of T_g. The cross-linked polymer chains relax causing the matrix to swell. As a result, the drug particles slowly diffuse out of the matrix. Hydroxypropyl methylcellulose (HPMC) and poly (ethylene oxide) (PEO) are regularly used for manufacturing such systems [10]. De la Torre *et al.* studied the release of two drugs, amoxicillin trihydrate, and amoxicillin sodium, from polyionic complexes of CS and poly (acrylic acid) (PAA) by observing the effect of electrostatic interactions between polymer-polymer and polymer-drugs [11]. Electrostatic polymer-polymer interactions occurred between the cationic groups of CS and the anionic groups of PAA. Amoxicillin trihydrate is a non-ionized drug, so its release was dictated only by polymer-polymer interactions. In case of the ionized drug amoxicillin sodium, additional electrostatic polymer-drug interactions also took place, as a result of which ionic interactions increased within the polymer network. This, in turn, hampered the drug release process.

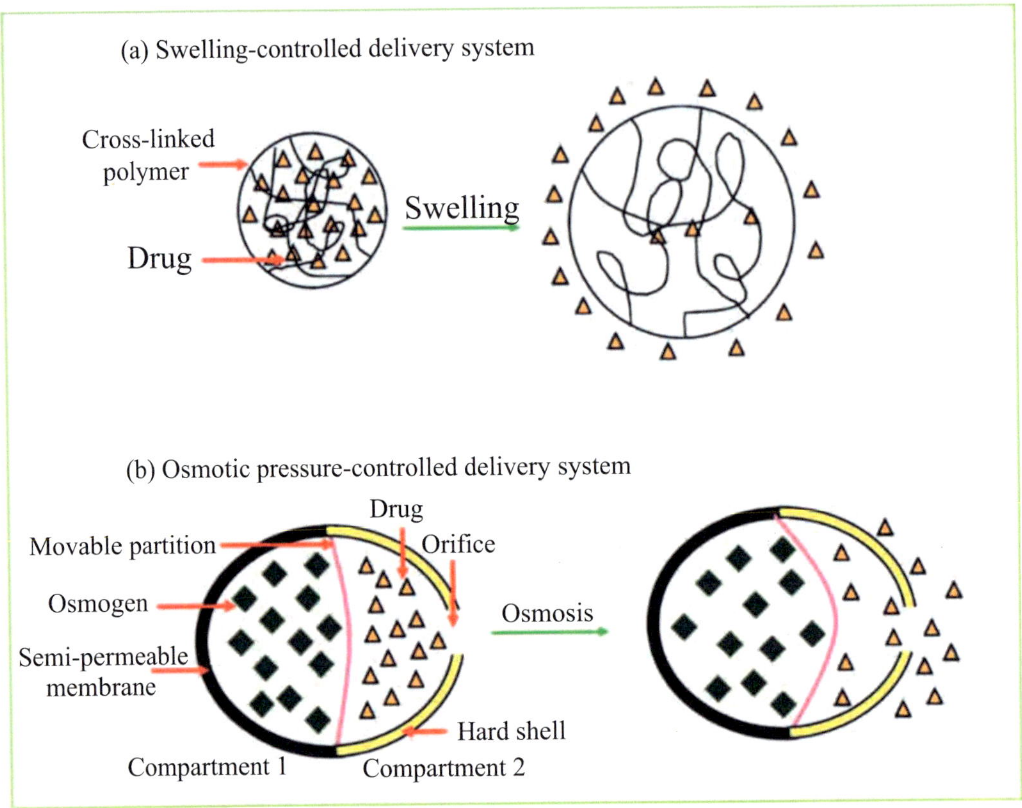

Fig. (2). Solvent-activated DDSs: **(a)** swelling-controlled and **(b)** osmotic pressure-controlled delivery systems.

Osmotic pressure-controlled DDS works on the principle of osmosis, *i.e.*, water can move from a higher concentration region to a lower concentration region across a semi-permeable membrane. A basic osmotic pump consists of two compartments. The first compartment is filled with an osmogen, coated with a polymeric membrane selectively permeable to only water. The osmogen creates osmotic pressure inside the DDS to fasten the release rate of a drug that has low solubility. The drug can itself be an osmogen, or osmogenic salt (sodium chloride and potassium chloride), and sugars (lactose, fructose, sucrose, and sorbitol) can be added to the formulation [12]. Cellulose acetate, cellulose triacetate, and cellulose esters are used as semi-permeable membranes. The second compartment is separated from the first one by a moving partition and contains drug particles. It is protected with a hard shell equipped with a delivery orifice. Water flows toward the second compartment, crossing the membrane and the first compartment, and creates pressure inside it. This forces the drugs to exit out of the DDS through the orifice as a solution or a suspension. Verma and Garg developed an extended-release formulation of glipizide, and studied how different formulation factors like pore former (PVP), membrane weight gain and solubility modifier (tromethamine) affected the drug release process [13]. The release of glipizide was directly proportional to the initial PVP level. PVP helped in forming more pores in the membrane to enable faster drug release. They also found that the release of glipizide was inversely related to the membrane weight gain. Solubility modifiers were used as alkalizing agents to raise the microenvironmental pH of the tablet's core above the drug's pKa value to increase its solubility. As a consequence, the presence of tromethamine facilitated the release of glipizide from the DDS.

Chemically Controlled DDS

In chemically controlled DDSs, polymers undergo a change in their chemical structure on exposure to the microenvironment inside the body, prompting the release of the drug [14]. Chemically controlled DDSs are categorized into drug-polymer dispersion systems and drug-polymer conjugate systems. In drug-polymer dispersion DDSs, biodegradable or bioerodible polymers, possessing hydrolytically or enzymatically labile bonds, are used to dissolve or uniformly disperse the drug molecules. They degrade or erode in response to the adjacent biological milieu and release the drug into its surroundings (Fig. **3**). The term biodegradation is used when a decrease in the molecular weight of the polymer is observed due to chain cleavage. Degradation of the polymer continues to occur long after the drug is released. On the other hand, in bioerosion, a decrease in the mass of the polymer occurs. Erosion of the polymer and release of the drug take place in a parallel time scale.

Depending on how erosion occurs, it can be divided into surface or bulk erosion. In surface erosion, the dissolution of polymers takes place from the surface or they breakdown from the surface, while in bulk erosion, the breakdown of polymers occurs in the bulk. PGA, PLA, PLGA, polyaminoacids, poly (ortho esters) and polyanhydrides are commonly used bioerodible/biodegradable polymers. Sundararaj *et al.* developed a bioerodible system (surface) from cellulose acetate phthalate (CAP) and Pluronic F-127 (P) [15]. The polymers were joined by hydrogen bonds in an aprotic solvent to form CAPP. Four different drugs (metronidazole, doxycycline, ketoprofen and simvastatin) and a model protein (lysozyme) were loaded in the CAPP films, and their *in vitro* release was observed both from single-layered and multi-layered DDSs. During the entire period of erosion, an individual drug or the protein-loaded single-layered CAPP exhibited sustained release. Four-layered CAPP films with alternating drug-loaded and blank layers caused its intermittent release. In case all four drugs were used in a sequential manner with alternating blank layers, the drugs were also sequentially released according to their loading position.

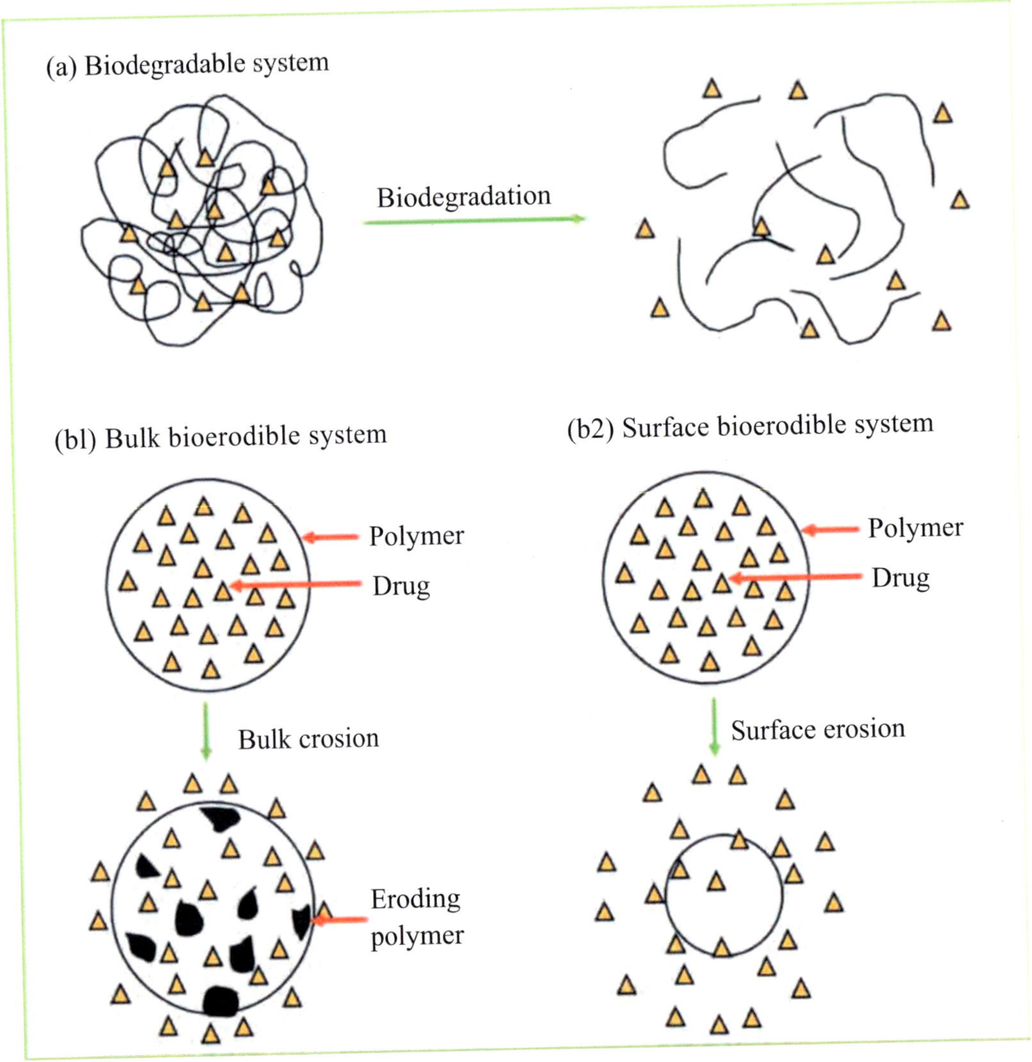

Fig. (3). Chemically controlled DDSs: **(a)** biodegradable and **(b)** bio erodible drug-polymer dispersion systems.

In drug-polymer conjugate DDSs, the drug is chemically bound to a polymer backbone, mostly through a spacer molecule, by covalent bonding, ionic bonding or any weak secondary molecular force. The drug is ultimately released *via* hydrolytic or enzymatic cleavage of the polymer-drug bond (Fig. **4**). Polymers used in such systems contain biodegradable or hydrolysable chemical linkages, examples of which are starch, chitosan, polysaccharides, polylysine, poly (glutamic acid), polyphosphazene, *etc*. For example, Oledzka *et al.* reported macromolecular ampicillin-branched polymer conjugates based on ester bonding [16]. The branched polymers, 6AR/PCL and 6AR/PLLA were composed of biodegradable poly (ε-caprolactone) (PCL) and poly (L-lactide) (PLLA) incorporated with natural arginine-6-oligomer (6AR), and were synthesized using the N, N'-dicyclohexylcarbodiimide (DCC)/ 4-dimethylaminopyridine (DMAP) coupling technique. 6AR was used as an initiator in ring opening polymerization (ROP) of the cyclic esters (ε-caprolactone and L-lactide) to enable them to form branched polymer matrices for covalent conjugation with the drug. After 21 days of incubation, the amount of ampicillin released from 6AR/PLLA/ampicillin conjugate (60%) was higher than that from 6AR/PCL/ampicillin (14%).

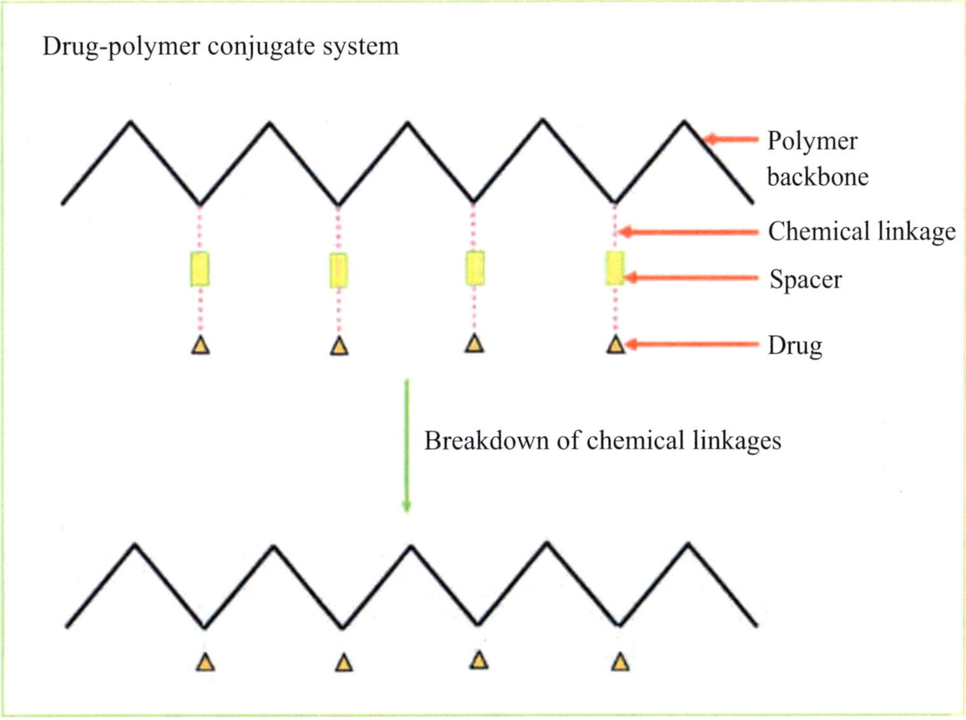

Fig. (4). Chemically controlled DDSs: drug-polymer conjugate system.

Externally Triggered DDS

Drug release at the desired location can be triggered by applying external stimuli like pH, temperature, light, magnetic and electric fields, chemical and biochemical agents, *etc* [17]. For this purpose, smart polymers are used that can undergo physical or chemical transformation (up to a critical limit) in response to an external stimulus. Once the stimulus is removed, the smart polymers have the ability to return to their original form. Physical stimuli (temperature, light, magnetic and electrical fields, *etc.*) directly modulate the polymer/solvent system's energy level, while chemical stimuli (pH, chemical agents, *etc*,) alter polymer-solvent interactions or molecular interactions between the polymer chains. Some of the common external stimuli are discussed below.

pH-Responsive Systems

pH-responsive polymers are generally polyelectrolytes containing a large number of weak ionizable acidic (carboxylic acids, sulfonic acids, *etc.*) or basic (ammonium salts) moieties attached to a hydrophobic backbone. They can accept or donate protons in reaction to changes in the pH of the solution surrounding the DDS [18]. Different parts of the body have different pH values (*e.g.*, blood (7.35-7.45), colon (7.0-7.5), stomach (1.0-3.0)). These changes in pH within the body can induce a reaction, like swelling or collapse of the polymer. For example, Bhattacharyya *et al.* prepared beads of a semi-interpenetrating network of PU-alginate for the delivery of bovine serum albumin (BSA) protein [19]. The swelling behaviour and release of BSA in different pH (1.2 and 7.4) were investigated. They found that at pH 1.2, the degree of swelling of the beads and release of encapsulated BSA in the stomach was minimal, but at pH 7.4, the degree of swelling of the beads increased by 1080% accompanied by practically complete release of BSA in the intestine.

Temperature Responsive Systems

The normal human body temperature is 37°C but it often changes in the presence of foreign microorganisms. Thus, temperature difference can be used to trigger drug release at the target location. Thermoresponsive polymers maintain a very delicate hydrophobic–hydrophilic balance with the aqueous medium, and a small deviation in temperature near its critical solution temperature (CST) (the temperature at which phase transition happens) can cause a sudden change in the polymer's physical or chemical properties. These are divided into two classes depending on whether they display a lower critical solution temperature (LCST) or an upper critical solution temperature (UCST). LCST-type polymers are soluble below LCST but undergo a sharp transition from coil form to globule form above their LCST. This leads to a rise in entropy, as a result of which the polymers drop out of the solution. LCST-type polymers are more common. Poly (N-isopropylacrylamide) (PNIPAAm) is a prime example of having LCST onset very close to human body temperature (32°C). Conversely, UCST polymers interact properly with the solvent only above the UCST [20]. Chung *et al.*

prepared a temperature-sensitive polymeric micelle using block copolymers of (PNIPAAm-b-butyl methacrylate) (PNIPAAm-PBMA) [21]. The core of the micelle was composed of self-aggregates of PBMA segments and the shell consisted of thermoresponsive PNIPAAm chains. An anti-cancer drug, Adriamycin, was successfully loaded inside the PBMA core. Below LCST, PNIPAAm aided in the stabilization of the system. But on heating the system, PNIPAAm underwent reversible structural deformation and initiated drug release above LCST.

Magnetically Controlled Systems

Magnetic nanoparticles (MNPs) are nanomaterials containing magnetic elements, which have found tremendous application in localized drug delivery, particularly for cancer patients. MNPs act as drug carriers, which can be directed to the tumour site by an external magnetic field. Polymer-MNP composites incorporate inorganic MNPs inside polymer coatings to stabilize as well as shield them from environmental degradation. Dextran, polysaccharides, PEG, *etc*. are often used for this purpose. A composite of MNPs and smart polymers is preferred since their properties can be tuned to respond to different stimuli (temperature, pH) in addition to the magnetic field.

MNPs in Drug Delivery

A magnetic field has nearly zero effect on the biological tissue, thus MNPs loaded with drugs can be utilized for targeted drug delivery. Currently, the properties of MNPs are harvested in two ways: MNPs improve the accumulation and release of drugs at the target site *via* magnetic targeting, and the magnetic field is used to heat up MNPs to induce hyperthermia in the diseased region. In the second case, the heat generated can also trigger drug release from thermoresponsive polymer carriers.

Magnetic Targeting

For magnetic targeting, the drug and MNPs are entrapped inside a drug carrier, and an external magnetic field is used to guide it inside the body to reach the target site where it accumulates. Approaches used for targeting of MNPs are often non-specific (passive targeting) [22, 23]. One of them is based on the enhanced permeability and retention (EPR) effect exhibited by most tumours suffering from a leaky vasculature (Fig. **5**). EPR effect increases the leakage of MNPs and polymers from the arteries and helps in their accumulation in the tumour. Another passive targeting approach involves taking up MNPs by the cells of the reticuloendothelial system (RES), which includes phagocytes, bone marrow cells, macrophages, *etc*. However, for proper accumulation of MNPs, certain things have to be kept in mind. Firstly, they must circulate long inside the body (blood) before being removed by the RES. Secondly, their size (20-500 nm) should be such that they are not too large to pass through endothelial cells of the tumour vasculature nor are they too small to escape from inside the tumour. Thirdly, during circulation, the MNPs (non-specifically) interact with the RES. So, MNPs have to be coated with hydrophilic polymers (like dextran, PEG or poly (N-(--hydroxypropyl) methacrylamide) (PHPMA)) to prevent rapid uptake by phagocytosis and the RES. The hydrophilic coating also extends the circulation time in the blood. Nevertheless, the EPR effect is fairly universal, so the DDSs can be used in the treatment of a variety of tumours by passive targeting.

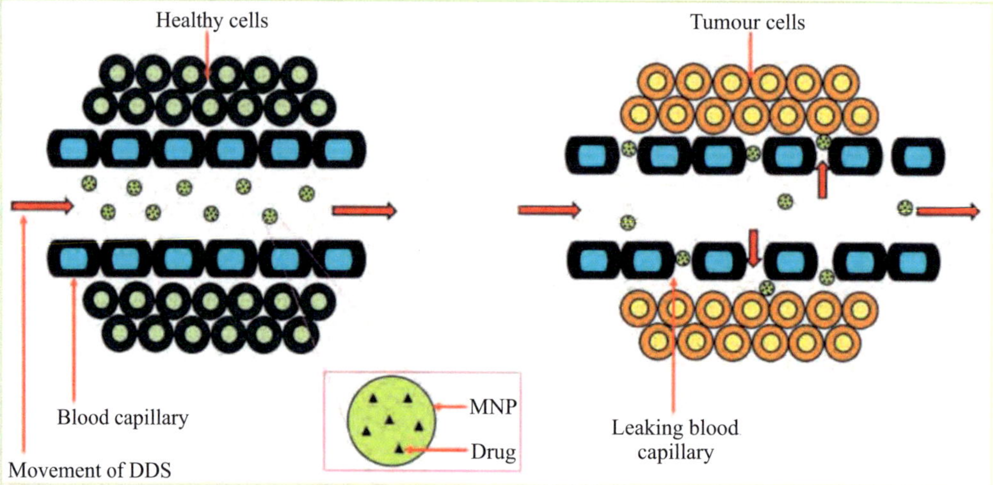

Fig. (5). Passively targeting DDSs based on EPR effect.

When it comes to selective targeting, an active targeting approach is better suited. This is based on receptor-ligand interactions (Fig. **6**). The concentration of certain surface receptors (like cell membrane proteins, glycoproteins and lipoproteins) is higher in the tumour cells than in healthy cells. DDSs can be conjugated with ligands that bind to these receptors, and safely deliver them to the target site bypassing the healthy cells. An MNP-based DDS for active targeting is essentially composed of the drug-loaded carrier with an additional targeting ligand. An example of a receptor is galactose found in liver tumour cells. Seymour *et al.* prepared hepatocyte galactose-targeted HPMA copolymer conjugated with doxorubicin (drug) and galactosamine (ligand) to facilitate specific binding to parenchymal liver cells [24].

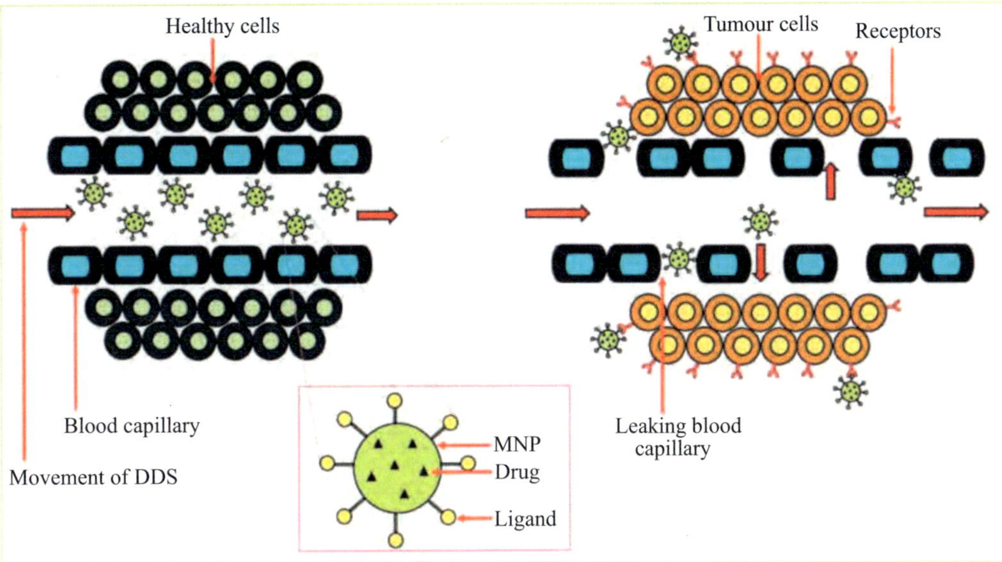

Fig. (6). Actively targeting DDSs based on ligand-receptor interactions.

The DDS is mostly injected inside the body into the circulatory system as a colloidal suspension (ferrofluid). Then a strong magnet, like Nd–Fe–B, is used externally over the target site to generate a high-gradient magnetic field. Once the magnetic forces overcome the force of blood in arteries (10 cm s^{-1}) or capillaries (0.05 cm s^{-1}), the DDSs accumulate at the tumour site, and the drug is subsequently released [25].

Magnetically-Induced Hyperthermia

In this scenario, diseased tissues are heated by MNPs (Fig. **7**). MNPs are exposed to an alternating magnetic field (AMF) that produces a directional force on each individual MNP [9]. Due to oscillation at high frequency, the average force on the MNP becomes zero. The energy of the oscillation is transformed into heat, which raises the temperature of the MNPs and subsequently, adjacent cells. This process is called magnetic hyperthermia (MH). MH is non-invasive and produces heat locally to damage the tumour cells, without affecting nearby health cells. But for that to happen, MNPs should have a narrow size distribution, and should be able to produce a sufficient temperature rise (43-46 °C). For instance, Clerc *et al.* developed a DDS for targeted magnetic intra-lysosomal hyperthermia formulated with magnetic iron oxide nanoparticles coated with PEG-amine and grafted with the ligand gastrin [26]. The receptor for gastrin (CCK2R) is overexpressed in the lysosomes of cancer cells. Upon application of high-frequency AMF, the heat generated by MNPs enhanced the production of reactive oxygen species (ROS) within lysosomes through the Fenton reaction catalysed by the iron ions. ROS caused lysosome membrane permeabilization (LMP) *via* lipid peroxidation of the lysosome membrane, leading to the destruction of tumour cells.

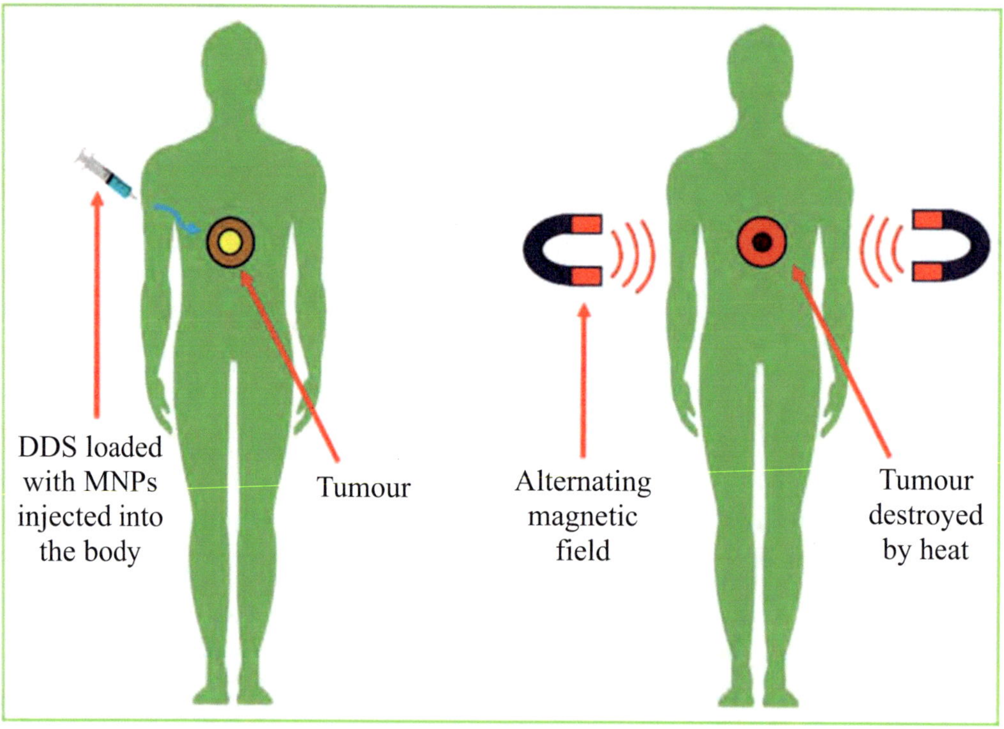

Fig. (7). Magnetic hyperthermia.

Superparamagnetic Iron Oxide Nanoparticles (SPIONs)

SPIONs are the most predominantly used MNPs in DDSs. SPIONs are small particles of iron oxide (magnetite (Fe_3O_4), maghemite (γ-Fe_2O_3) or hematite (α-Fe_2O_3)) with hydrodynamic radii between 20-150 nm [27]. Amongst them, magnetite is largely chosen since iron exists as Fe^{2+} and Fe^{3+} in the ratio 1:2, and Fe^{2+} acts as a catalyst in the Fenton reaction leading to the production of ROS for killing tumour cells. The oxidation state of iron, on the other hand, is only +3 in maghemite and hematite, which is not very useful for targeted drug delivery. The main advantage of SPIONs lies in their superparamagnetic behaviour, *i.e.*, these MNPs have no hysteresis and zero coercivity due to thermal fluctuations. Bulk magnetite behaves as ferrimagnets, *i.e.*, they show residual magnetism even after the removal of an external magnetic field. But when their size becomes less than 20 nm, then they act as a single magnetic domain exhibiting high magnetic susceptibility. When an external magnetic field is applied, the SPIONs are magnetized up to their saturation magnetization, and once the field is removed, they show negligible residual magnetism at room temperature (Fig. **8**) [28]. Thus, superparamagnetism plays a vital role in dragging the DDSs to the target site.

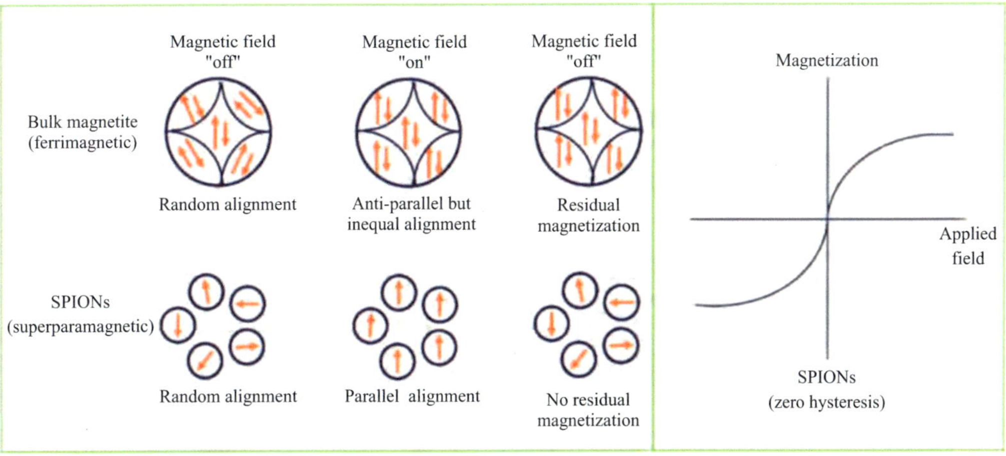

Fig. (8). Magnetic behaviour of bulk magnetite and SPIONs.

Effect of Size

The size of SPIONs plays a crucial role in DDSs. Smaller MNPs have higher magnetic moments per unit volume, so they respond more effectively to AMF. Also, they have higher surface area to volume ratio, so they can have higher drug payload. However, if their size becomes too small, they tend to agglomerate when the AMF is removed. It also becomes challenging to keep them at the target site and resist the drag of blood flow in the absence of AMF. Conversely, larger-sized MNPs are more stable and tend to aggregate less. However, they can load a lesser number of drug particles onto them [29]. Hence, the size of the SPIONs needs to be regulated properly during the synthesis process according to their intended application.

Smart Polymer/SPION Nanocomposites

Although the application of SPIONs in therapeutics seems very lucrative, it is also riddled with a few problems. Firstly, magnetic heating requires enormously high concentrations of SPIONs (10–100 mg mL^{-1}) [30]. At such concentrations, naked SPIONs can hinder cell growth *in vivo*, thus disrupting the cell cycle. High concentration and repeated exposure to SPIONs can also induce cytotoxicity (iron poisoning) and oxidative stress if they decay into free iron. Secondly, naked SPIONs have the tendency to aggregate with serum proteins. Thirdly, they are prone to quick elimination from circulation by the RES [31]. To overcome these critical issues, DDS formulations incorporate SPIONs within biocompatible and biodegradable polymers. The polymer coating serves many functions: it shields the SPIONs from environmental degradation, prevents aggregate formation and also acts as binding sites that can be functionalised to attach drugs or other molecules as required. The attachment between the SPION/polymer nanocomposite and the drug molecules can be covalent bonding, adsorption, entrapment, *etc.*

Magnetic Core-Shell Structure

A core-shell structure has two discrete parts: the inner part is called the core and the outer layer is called the shell. The shell completely encompasses the core. Both the core and the shell can perform individual jobs. For example, Tansık *et al.* synthesized PLGA-coated SPION core-shell nanocomposites using oil in water emulsion technique. They loaded the nanocomposite with an anti-cancer drug, doxorubicin by co-encapsulating it inside PLGA along with SPIONs, and studied its cytotoxic effect on the MCF-7 cell line. Their analyses showed that the DDS caused significant cell death with a dosage of 16-250 $\mu g\ mL^{-1}$ [32]. In another experiment, Hoang Thi *et al.* prepared a core-shell structure of SPION core coated with heparin-poloxamer (HP) copolymer by the co-precipitation method. Doxorubicin was successfully encapsulated inside the polymeric shell. The DDS demonstrated controlled drug release upon biodegradation of HP at a steady rate up to 120 h [33]. While the core in DDSs mostly comprises SPIONs, polymeric materials used as shells include PVA, PEG, PLGA, dextran and polysaccharides, *etc.* (Fig. **9**). These types of drug carriers are multifunctional: the core can help in drug circulation by providing magnetic guidance, while the shell can promote controlled drug release *via* biodegradation of the polymer.

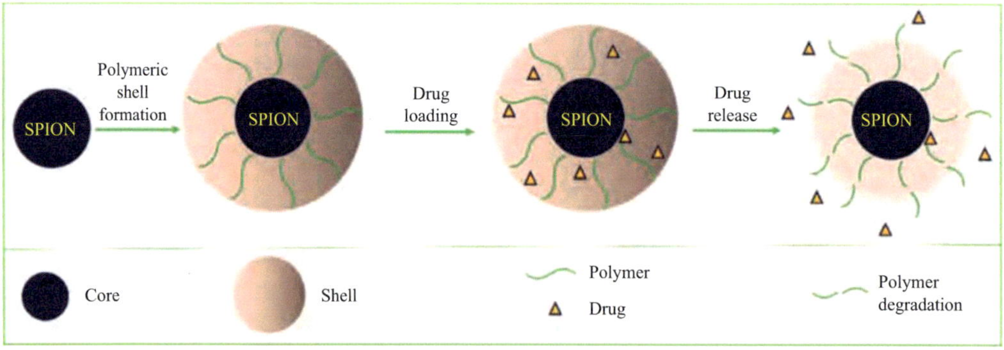

Fig. (9). Drug release from magnetic core-shell structure.

Thermally-Responsive Magnetic Carriers

Conversely, the polymer can also act as the core with SPIONs acting as the shell. For example, Oka *et al.* prepared core-shell nanocomposites of an assembly of

SPIONs-coated polyhydroxyalkanoate (PHA) particles using a modified emulsification solvent diffusion method. They loaded the DDS with a model drug, pyrene by simply adding it during the nanocomposite formation stage. They theorized that it was possible to release the drug by two mechanisms: continued release when PHA underwent biodegradation and quick release when PHA was melted by the heat generated by SPIONs in the presence of an AMF. Although the second mechanism was expected, it was not experimentally proven in that paper [34]. These types of formulations designed by Oka and his group are known as thermally responsive magnetic DDSs, wherein an AMF is used to trigger drug release due to heat generated by SPIONs. The effect is termed as hot spot effect. A major advantage of such a design is that drug release can be achieved with relatively lower concentrations of SPIONs. In another study, Wadajkar *et al.* successfully synthesized thermally-responsive poly (N-isopropylacrylamid--acrylamide-allylamine) (PNIPAAm-AAm-AH) coated SPIONs for active targeting of prostate cancer by free radical polymerization route. The polymeric coating had amine functional groups that were used to conjugate R11 peptides to the DDS *via* carbodiimide chemistry, specifically for receptor-mediated targeting of prostate cancer cells. The LCST of the composite was 40 °C. Below LCST, the DDS showed a hydrophilic character. But above LCST, the polymer structure collapsed causing dehydration of the polymer chain. It became hydrophobic and led to drug release at the target site. The SPIONs, in this case, served dual functions: helped in guiding the core-shell nanocomposites to the tumour site and also act as contrast agents in magnetic resonance imaging (MRI) to observe the accumulation of drug particles *in vivo* [35].

Multistimuli-Responsive Magnetic Carriers

The synthesis of the majority of the core-shell structures mentioned in the previous paragraphs involved a two-step process: the core is synthesized in the first step and then the polymer shell is coated on the core by a polymerization method. However, it is possible to have more than one shell of smart polymer with each shell displaying individual stimulus-responsive behaviour [36]. For example, Zeng *et al.* prepared a core-shell architecture of SPIONs core shelled with a layer of dual responsive (reduction/pH) PMMA. A second layer of thermal-responsive PNIPAAm was then added as a gatekeeper on the surface, and finally, the doxorubicin hydrochloride drug was loaded onto it. The core was synthesized by co-precipitation of Fe^{2+} and Fe^{3+} ions in an alkali solution. The inner shell of PMMA was made by a distillation precipitation polymerization process using N, N-bis(acryloyl)cystamine (BACy) as a cross-linker. BACy contained disulfide bonds that were easily disrupted in the presence of dithiothreitol (DTT) or glutathione (GSH) reducing agents. The outer shell of PNIPAAm was then made by a two-stage distillation precipitation polymerization route exploiting hydrogen bonding between PMAA and NIPAM monomers, and also by using cross-linker N, N'-methylenebisacrylamide (MBA) during the second stage of polymerization. The composite was loaded with Doxorubicin hydrochloride, and its release profile was examined in response to changes in pH, reducing agents and temperature. They found that the drug release at 37 °C after 24 h was much faster at pH 5.0 (25%) as compared to pH 7.4 (11%). Also, 10 mM GSH (88%) aided in release much more efficiently than 10 mM DTT (26%) [37].

CONCLUSION AND FUTURE PERSPECTIVE

To formulate an efficient DDS, the selection of a suitable smart polymer is very important. Polymeric composites play a significant role in therapeutics due to the presence of functional moieties that help in binding or adsorbing drugs and other molecules. Once loaded, drugs can be released at the intended target by the combined effects of polymer degradation and drug diffusion under specific conditions. Drug release can be controlled by the presence of a lot of internal as well as external stimuli-responsive smart polymers. One of the stimuli is the use of an external AMF to treat cancer cells using polymer/SPION/drug formulations. These formulations are multi-functional: (i) they provide magnetic guidance during circulation, (ii) trigger drug release at tumour sites and (iii) kill diseased cells *via* magnetic hyperthermia. They can also be combined with other stimuli (*e.g.*, temperature, pH, redox) to design multistimuli-responsive platforms for higher therapeutic efficacy.

The mechanism of how a DDS interacts with a particular tissue or cells still needs to be studied in depth in order to design the most suitable formulation so that the DDS can be properly translated from the experimental stage to commercial production. With the incorporation of SPIONs into DDSs, some key parameters like suitable size and uniform size distribution for longer circulation, uniform response to external AMF, and biodegradability need to be better controlled for effective biomedical applications. In the current therapeutic landscape, the treatment of advanced stage cancer and crossing the blood-brain barrier continue to pose problem areas for researchers. In the case of tumour targeting, work has to be done on delivering high-potency drugs to specific sites. Also, transporting drugs into the deeper parts of the brain to treat psychiatric and neurological conditions without damaging the blood-brain barrier is an ongoing matter. In the future, the treatment of emerging lifestyle diseases will also require advanced smart drug carriers for rapid and regulated delivery of their payload. With these issues in mind, scientists will need to have an interdisciplinary and open-minded style towards designing DDSs.

REFERENCES

[1] Ji, T.; Kohane, D.S. Nanoscale systems for local drug delivery. *Nano Today,* **2019**, *28*, 100765. [http://dx.doi.org/10.1016/j.nantod.2019.100765] [PMID: 32831899]

[2] Vargason, A.M.; Anselmo, A.C.; Mitragotri, S. The evolution of commercial drug delivery technologies. *Nat. Biomed. Eng.,* **2021**, *5*(9), 951-967. [http://dx.doi.org/10.1038/s41551-021-00698-w] [PMID: 33795852]

[3] Liechty, W.B.; Kryscio, D.R.; Slaughter, B.V.; Peppas, N.A. Polymers for drug delivery systems. *Annu. Rev. Chem. Biomol. Eng.,* **2010**, *1*(1), 149-173. [http://dx.doi.org/10.1146/annurev-chembioeng-073009-100847] [PMID: 22432577]

[4] Adepu, S.; Ramakrishna, S. Controlled drug delivery systems: current status and future directions. *Molecules,* **2021**, *26*(19), 5905. [http://dx.doi.org/10.3390/molecules26195905] [PMID: 34641447]

[5] Lee, P.I.; Good, W.R. Overview of controlled-release drug delivery. *ACS Symposium Series,* **1987**, *Vol. 348*, pp. 1-13. [http://dx.doi.org/10.1021/bk-1987-0348.ch001]

[6] Kim, J.H.; Kim, J.Y.; Lee, Y.M.; Kim, K.Y. Controlled release of riboflavin and insulin through crosslinked poly(vinyl alcohol)/chitosan blend membrane. *J. Appl. Polym. Sci.,* **1992**, *44*(10), 1823-1828. [http://dx.doi.org/10.1002/app.1992.070441015]

[7] Qureshi, D.; Nayak, A.K.; Kim, D.; Maji, S.; Anis, A.; Mohanty, B.; Pal, K. Polysaccharide-based polymeric gels as drug delivery vehicles.*Advances and Challenges in Pharmaceutical Technology*; Nayak, A.K.; Pal, K.; Banerjee, I.; Maji, S.; Nanda, U., Eds.; Academic Press, **2021**, pp. 283-325. [http://dx.doi.org/10.1016/B978-0-12-820043-8.00013-X]

[8] Majid Khan, G.; Zhu, J.B. Studies on drug release kinetics from ibuprofenâ€"carbomer hydrophilic matrix tablets: influence of co-excipients on release rate of the drug. *J. Control. Release,* **1999**, *57*(2), 197-203. [http://dx.doi.org/10.1016/S0168-3659(98)00122-9] [PMID: 9971903]

[9] Coelho, J.F.; Ferreira, P.C.; Alves, P.; Cordeiro, R.; Fonseca, A.C.; Góis, J.R.; Gil, M.H. Drug delivery systems: Advanced technologies potentially applicable in personalized treatments. *EPMA J.,* **2010**, *1*(1), 164-209. [http://dx.doi.org/10.1007/s13167-010-0001-x] [PMID: 23199049]

[10] Colombo, P.; Bettini, R.; Santi, P.; Peppas, N.A. Swellable matrices for controlled drug delivery: gel-layer behaviour, mechanisms and optimal performance. *Pharm. Sci. Technol. Today,* **2000**, *3*(6), 198-204. [http://dx.doi.org/10.1016/S1461-5347(00)00269-8] [PMID: 10840390]

[11] Torre, P.; Enobakhare, Y.; Torrado, G.; Torrado, S. Release of amoxicillin from polyionic complexes of chitosan and poly(acrylic acid). study of polymer/polymer and polymer/drug interactions within the network structure. *Biomaterials,* **2003**, *24*(8), 1499-1506. [http://dx.doi.org/10.1016/S0142-9612(02)00512-4] [PMID: 12527291]

[12] Almoshari, Y. Osmotic pump drug delivery systems—a comprehensive review. *Pharmaceuticals (Basel),* **2022**, *15*(11), 1430. [http://dx.doi.org/10.3390/ph15111430] [PMID: 36422560]

[13] Verma, R.K.; Garg, S. Development and evaluation of osmotically controlled oral drug delivery system of glipizide. *Eur. J. Pharm. Biopharm.,* **2004**, *57*(3), 513-525. [http://dx.doi.org/10.1016/j.ejpb.2004.02.003] [PMID: 15093601]

[14] Bajpai, A.K.; Shukla, S.K.; Bhanu, S.; Kankane, S. Responsive polymers in controlled drug delivery. *Prog. Polym. Sci.,* **2008**, *33*(11), 1088-1118. [http://dx.doi.org/10.1016/j.progpolymsci.2008.07.005]

[15] Sundararaj, S.C.; Thomas, M.V.; Dziubla, T.D.; Puleo, D.A. Bioerodible system for sequential release of multiple drugs. *Acta Biomater.,* **2014**, *10*(1), 115-125. [http://dx.doi.org/10.1016/j.actbio.2013.09.031] [PMID: 24096151]

[16] Oledzka, E.; Sobczak, M.; Nalecz-Jawecki, G.; Skrzypczak, A.; Kolodziejski, W. Ampicillin-ester bonded branched polymers: characterization, cyto-, genotoxicity and controlled drug-release behaviour. *Molecules,* **2014**, *19*(6), 7543-7556. [http://dx.doi.org/10.3390/molecules19067543] [PMID: 24914899]

[17] Wang, Y.; Kohane, D.S. External triggering and triggered targeting strategies for drug delivery. *Nat. Rev. Mater.,* **2017**, *2*(6), 17020. [http://dx.doi.org/10.1038/natrevmats.2017.20]

[18] Balamurali, V.; .Pramodkuma, T.M.; Srujana, N.; Venkatesh, M.P.; Gupta, N.V.; Krishna, K.L.; Gangadhara, H.V. pH sensitive drug delivery systems: a review. *American Journal of Drug Discovery and Development,* **2010**, *1*(1), 24-48. [http://dx.doi.org/10.3923/ajdd.2011.24.48]

[19] Bhattacharyya, A.; Mukhopadhyay, P.; Kundu, P.P. Synthesis of a novel pH-sensitive polyurethane-alginate blend with poly(ethylene terephthalate) waste for the oral delivery of protein. *J. Appl. Polym. Sci.,* **2014**, *131*(16), app.40650. [http://dx.doi.org/10.1002/app.40650]

[20] Pasparakis, G.; Tsitsilianis, C. LCST polymers: Thermoresponsive nanostructured assemblies towards bioapplications. *Polymer (Guildf.),* **2020**, *211*, 123146. [http://dx.doi.org/10.1016/j.polymer.2020.123146]

[21] Chung, J.E.; Yokoyama, M.; Yamato, M.; Aoyagi, T.; Sakurai, Y.; Okano, T. Thermo-responsive drug delivery from polymeric micelles constructed using block copolymers of poly(N-isopropylacrylamide) and poly(butylmethacrylate). *J. Control. Release,* **1999**, *62*(1-2), 115-127. [http://dx.doi.org/10.1016/S0168-3659(99)00029-2] [PMID: 10518643]

[22] kianfar, E. Magnetic nanoparticles in targeted drug delivery: a review. *J. Supercond. Nov. Magn.,* **2021**, *34*(7), 1709-1735. [http://dx.doi.org/10.1007/s10948-021-05932-9]

[23] Ulbrich, K.; Holá, K.; Šubr, V.; Bakandritsos, A.; Tuček, J.; Zbořil, R. Targeted drug delivery with polymers and magnetic nanoparticles: covalent and noncovalent approaches, release control, and clinical studies. *Chem. Rev.,* **2016**, *116*(9), 5338-5431. [http://dx.doi.org/10.1021/acs.chemrev.5b00589] [PMID: 27109701]

[24] Seymour, L.W.; Ulbrich, K.; Wedge, S.R.; Hume, I.C.; Strohalm, J.; Duncan, R. N-(--Hydroxypropyl)methacrylamide copolymers targeted to the hepatocyte galactose-receptor: pharmacokinetics in DBA2 mice. *Br. J. Cancer,* **1991**, *63*(6), 859-866. [http://dx.doi.org/10.1038/bjc.1991.190] [PMID: 1648946]

[25] Mahmoudi, M.; Sant, S.; Wang, B.; Laurent, S.; Sen, T. Superparamagnetic iron oxide nanoparticles (SPIONs): Development, surface modification and applications in chemotherapy. *Adv. Drug Deliv. Rev.,* **2011**, *63*(1-2), 24-46. [http://dx.doi.org/10.1016/j.addr.2010.05.006] [PMID: 20685224]

[26] Clerc, P.; Jeanjean, P.; Hallali, N.; Gougeon, M.; Pipy, B.; Carrey, J.; Fourmy, D.; Gigoux, V. Targeted Magnetic Intra-Lysosomal Hyperthermia produces lysosomal reactive oxygen species and causes Caspase-1 dependent cell death. *J. Control. Release,* **2018**, *270*, 120-134. [http://dx.doi.org/10.1016/j.jconrel.2017.11.050] [PMID: 29203413]

[27] Dulińska-Litewka, J.; Łazarczyk, A.; Hałubiec, P.; Szafrański, O.; Karnas, K.; Karewicz, A. Superparamagnetic iron oxide nanoparticles-current and prospective medical applications. *Materials (Basel),* **2019**, *12*(4), 617. [http://dx.doi.org/10.3390/ma12040617] [PMID: 30791358]

[28] Wahajuddin, N.; Arora, S. Superparamagnetic iron oxide nanoparticles: magnetic nanoplatforms as drug carriers. *Int. J. Nanomedicine,* **2012**, *7*, 3445-3471. [http://dx.doi.org/10.2147/IJN.S30320] [PMID: 22848170]

[29] Nordin, A.; Ahmad, Z.; Husna, S.; Ilyas, R.; Azemi, A.; Ismail, N.; Nordin, M.; Ngadi, N.; Siti, N.; Nabgan, W.; Norfarhana, A.; Azami, M. The state of the art of natural polymer functionalized Fe_3O_4 magnetic nanoparticle composites for drug delivery applications: a review. *Gels,* **2023**, *9*(2), 121. [http://dx.doi.org/10.3390/gels9020121] [PMID: 36826291]

[30] Liu, J.F.; Jang, B.; Issadore, D.; Tsourkas, A. Use of magnetic fields and nanoparticles to trigger drug release and improve tumor targeting. *Wiley Interdiscip. Rev. Nanomed. Nanobiotechnol.,* **2019**, *11*(6), e1571. [http://dx.doi.org/10.1002/wnan.1571] [PMID: 31241251]

[31] Meyer, R.A.; Green, J.J. Biodegradable polymer iron oxide nanocomposites: the future of biocompatible magnetism. *Nanomedicine (Lond.),* **2015**, *10*(23), 3421-3425. [http://dx.doi.org/10.2217/nnm.15.165] [PMID: 26608843]

[32] Tansık, G.; Yakar, A.; Gündüz, U. Tailoring magnetic PLGA nanoparticles suitable for doxorubicin delivery. *J. Nanopart. Res.,* **2014**, *16*(1), 2171. [http://dx.doi.org/10.1007/s11051-013-2171-7]

[33] Hoang Thi, T.T.; Nguyen Tran, D.H.; Bach, L.G.; Vu-Quang, H.; Nguyen, D.C.; Park, K.D.; Nguyen, D.H. Functional magnetic core-shell system-based iron oxide nanoparticle coated with biocompatible copolymer for anticancer drug delivery. *Pharmaceutics,* **2019**, *11*(3), 120. [http://dx.doi.org/10.3390/pharmaceutics11030120] [PMID: 30875948]

[34] Oka, C.; Ushimaru, K.; Horiishi, N.; Tsuge, T.; Kitamoto, Y. Core–shell composite particles composed of biodegradable polymer particles and magnetic iron oxide nanoparticles for targeted drug delivery. *J. Magn. Magn. Mater.,* **2015**, *381*, 278-284. [http://dx.doi.org/10.1016/j.jmmm.2015.01.005]

[35] Wadajkar, A.S.; Menon, J.U.; Tsai, Y.S.; Gore, C.; Dobin, T.; Gandee, L.; Kangasniemi, K.; Takahashi, M.; Manandhar, B.; Ahn, J.M.; Hsieh, J.T.; Nguyen, K.T. Prostate cancer-specific thermo-responsive polymer-coated iron oxide nanoparticles. *Biomaterials,* **2013**, *34*(14), 3618-3625. [http://dx.doi.org/10.1016/j.biomaterials.2013.01.062] [PMID: 23419645]

[36] Karimi, M.; Sahandi Zangabad, P.; Ghasemi, A.; Amiri, M.; Bahrami, M.; Malekzad, H.; Ghahramanzadeh Asl, H.; Mahdieh, Z.; Bozorgomid, M.; Ghasemi, A.; Rahmani Taji Boyuk, M.R.; Hamblin, M.R. Temperature-responsive smart nanocarriers for delivery of therapeutic agents: applications and recent advances. *ACS Appl. Mater. Interfaces,* **2016**, *8*(33), 21107-21133. [http://dx.doi.org/10.1021/acsami.6b00371] [PMID: 27349465]

[37] Zeng, J.; Du, P.; Liu, L.; Li, J.; Tian, K.; Jia, X.; Zhao, X.; Liu, P. Superparamagnetic reduction/pH/temperature multistimuli-responsive nanoparticles for targeted and controlled antitumor drug delivery. *Mol. Pharm.,* **2015**, *12*(12), 4188-4199. [http://dx.doi.org/10.1021/acs.molpharmaceut.5b00342] [PMID: 26554495]

CHAPTER 4

Environmental Applications of Smart Polymer Composites

Khanindram Baruah[1] and **Nimisha Kashyap**[1,*]

[1] *Department of Chemistry, North-Eastern Hill University, Shillong, Meghalaya, India*

Abstract: The growing population and depletion of fossil fuels not only hamper industrial system, but also affect everyday life. Rapid urbanization, change in human lifestyle and continuous production of toxic chemicals are directly increasing the amount of chemical waste, air/water polluting substances, which are causing hazardous affect to human health for which no adequate solution is available. Thus, developing new technologies has been a priority for researchers to resolve such issues. Currently available conventional technologies need further development or replacement with newer technologies. Substantial research efforts put forward different methods among which the use of smart polymer materials to reduce environmental problems is observed to be one of the best. Intelligent polymer materials which are strategically designed to conquer multiple properties in diverse conditions such as pH, temperature, stress or electric field made their presence necessary. The response rate of these materials against different environmental stimuli can be regulated by simple functionalization, leading to significant industrial and domestic applications. In this chapter, the readers will get an overview of the using smart materials for environmental issues, and possibilities of exploring further contributions for solving environmental challenges.

Keywords: Antifouling, Catalysis, Composite material, Environment, Sensing, Self-healing, Smart polymer, Water harvest.

INTRODUCTION

The impact of growing modern civilization and rapid changes in human lifestyle concerns the government and the scientific community due to its harsh involvement towards environmental problems [1]. At various times, incidents like the Kalamazoo river oil spill and Huangdao oil pipeline explosion have shown disastrous moments when smart materials played a crucial role in overcoming that situation. Governments from different countries worldwide have already issued alerts regarding the impact and resolutions to solve these environmental problems.

* **Corresponding author Nimisha Kashyap:** Department of Chemistry, North-Eastern Hill University, Shillong, Meghalaya, India; E-mail: nimisha.kashyap27@gmail.com

Partha Pratim Saikia & Bhaskar Jyoti Saikia (Eds.)

Therefore, advancement of materials/methods for detecting, preventing, and remedying environmental damage is essential to socio-ecological health.

The materials that sense and respond to changes in their surroundings in a very precise and reproducible manner are called Smart Materials [2, 3]. They are considered environmentally significant due to their several environment-driven applications including catalysis chemistry, sensor chemistry, and chemistry relevant to decontamination methodology that is relevant to environmental problems [4 - 9]. The application in catalysis chemistry is of utmost importance because of the ecological concerns of producing chemical and hazardous wastes. Sensor chemistry is essential due to misusing toxic materials for specific purposes. Again, negligence, carelessness, and ignorance in handling carcinogenic and toxic chemicals result in polluting research/industrial facilities, water reservoirs, and the atmosphere. Researchers worldwide are engaged in identifying materials that can act swiftly, and could be implemented to reduce various types of waste produced in chemical processes, detect the presence and concentration of contaminants, and decontaminate fouled environments.

In **a** recent study, poly(vinyl chloride) (PVC) membrane-based sensors were utilized to investigate the presence of antipsychotic agents [10]. In this study, El-Ragehy *et al.* constructed modified electrodes using tetraphenylborate or tetrakis (4-chlorophenyl) borate for selective detection of fluphenazine and nortriptyline hydrochlorides *via* ion-pair formation. Another poly(vinyl) chloride (PVC) membrane electrode was prepared by Liu *et al.* [11]. This electrode comprised of pethidine–phosphotung is used for the determination of pethidine hydrochloride drug in injections and tablets. Besides polymer materials, nanocomposites prepared by incorporating nanomaterials into polymeric materials are also known for their novel applications. In this chapter, we have discussed briefly about the different environmental applications of smart materials. We hope that the readers will get an overview of the recent developments and future endeavours of smart materials for solving ecological concerns.

ENVIRONMENTAL APPLICATIONS OF SMART POLYMER MATERIALS

The amalgamation of environmental science with material science unlocked the door to overcome future challenges. The use of smart polymer materials for environmental benefits has been the focus of research and development for quite some time. The use of smart polymer materials involving catalysis chemistry, sensing chemistry, and decontamination methodology is especially relevant in addressing different environmental challenges [4, 5]. Furthermore, smart polymer materials are also utilized for self-healing, antifouling, water harvesting, *etc.*, some of which are discussed in this section.

Catalysis

Catalysis plays a significant role in synthetic laboratories and industrial processes. Catalysis is categorized into two types: heterogeneous (substrate and catalyst state are different) and homogenous catalysis (substrate and catalyst state are same). In the race to develop sensitive, and cost-effective, and reliable catalysts, smart polymer-based materials withdraw much attention due to their enhanced stability, uniform dispersion behavior and effortless functionalization properties [4]. For example, Xiao *et al.* studied the use of temperature/pH-responsive gold nanoparticle-incorporated hybrid nanogels for catalytic reduction of 4-nitrophenol. The catalytic efficiency was significantly affected by tuning the temperature, kinetics, and thermodynamics of the chemical reaction [12]. In another example, a microgel system composed of Nisopropylmethacrylamide and methacrylic acid (MAa) encapsulated with bimetallic (Ag/Ni) nanoparticles was used to catalyse the degradation of different azo dyes (Fig. **1a**) [13]. Temperature-responsive smart catalytic systems are also well known for their high catalytic performance at specific temperatures. A graphene-supported nanocatalyst $Au_{core}Pt_{shell}$ bimetallic (G-Au@Pt) was used for the catalytic reduction of 4-nitrophenol. In the presence of $NaBH_4$, the catalytic activity, kapp = $9.44x10^{-3}$ s^{-1} was higher than the previously reported catalysts (Fig. **1b**) [14]. Also, a temperature-dependent catalytic study revealed that the catalyst performance was linearly increased in the temperature range of 15-30 °C. However, the catalytic performance dramatically drops on increasing the temperature, calling it "switched off" (<35 °C).

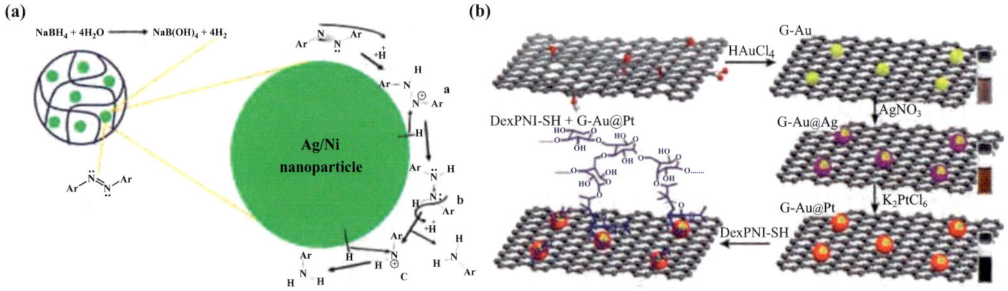

Fig. (1). (**a**) Plausible mechanism of catalytic degradation reaction of MOr using Ag/Ni-P(NMA-MAa)and $NaBH_4$ [13]. and (**b**) Synthesis of modified graphene sheets with $Au_{core}Pt_{shell}$ nanoparticles and attachment of the dextran-based smart polymer with thiol-ends [14].

Sensor

Devices capable of quickly and specifically responding to a certain type of analytes *via* significant chemical transformations are known as sensors. Typically, sensors are used for quantitative and/or qualitative analysis of the amount of analyte present in an unknown sample [15]. Chemical sensors are used mainly to detect/sense a wide variety of target species or analytes such as cations and anions, small organic molecules, gases, volatile organic compounds, explosives, chemical warfare agents, and bio-relevant species like proteins, glucose, and different disease markers. Smart polymer composite materials are widely used as sensor owing to their attractive physical characteristics like high selectivity, short response/recovery time, and wide range of working temperatures. For instance, molecularly imprinted polymers (MIPs) are used as template-based sensors for removing templates, with a substantial amount of cavities being exposed that act as receptor sites for the target. An earlier report by Peeters *et al.* demonstrated the use of a hybrid graphene oxide-MIP material with improved binding capacity and sensitivity [16]. In this work, a hybrid graphene oxide-MIP material was employed to detect histamine. A heat transfer method was used which requires two thermocouples with an adjustable heat source. As an outcome, a very low detection limit (25 nM) was observed concluding the promising applications of MIPs for binding and recognition. In another work, Wan and his group synthesized aenantiomeric phenylalanine incorporated MIP silica core-shell microparticle. This material was prepared using RAFT polymerization technique and shows an intense emission peak of around 520 nm in the presence of L-phenylalanine [17]. They mentioned that the system could be used in both pharmaceuticals as well as simple phase-transfer processes. Valderrey and coworkers described a fluorescent MIP that was used for selective recognition of enoxacin with a low detection limit [18] (Fig. **2**).

They prepared the MIP probe initially by functionalizing SiO_2 particles with 3-aminopropyltriethoxysilane followed by coupling with 4-cyano-4-(thiobenzoylthio)pentanoicacid as an RAFT agent. Later, an equimolar mixture of the MIP and a co-monomer along with ethylene glycol dimethacrylate was used to prepare an adequate template-dye complex

Advanced Oil/Water Separation

Oil/water mixture generated from petrochemical, food, textile, leather, and steel industries pose a serious threat to the environment and living species [19]. Traditional techniques used for the separation of emulsified oil/water mixtures are inefficient and a challenging task. Thus, bringing in novel, efficient materials /techniques for advanced oil/water separation attracts great attention. Pioneering work by Zhang and his coworkers introduced a very inexpensive and facile technique to fabricate a superhydrophobic- superoleophilic PVDF membrane [20]. The PVDF membrane demonstrated a remarkable ability to separate both micrometer and nanometer-sized water-in-oil emulsions. Similarly, Gao *et al.* fabricated porous nitrocellulose membranes and they showed impressive superoleophobicity and effective oil/water separation [21]. Recently, Jiang *et al.* developed ultralight nanofibrous aerogels (NFAs) with unique wetting properties showing great potential in oily wastewater treatment [22]. The obtained NFAs have a self-assembled hierarchical structure with an ultralow apparent density and they possess high hydrophobicity and superoleophilicity, demonstrating superabsorbent properties in separating oil from layered and emulsified oil–water mixtures.

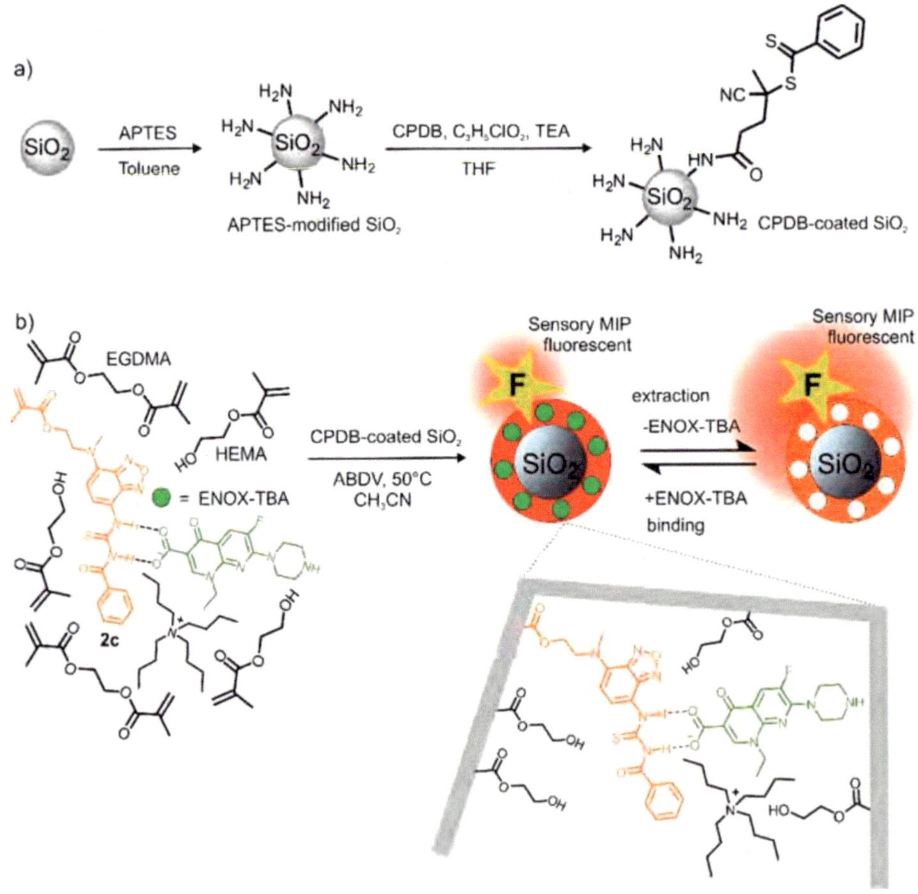

Fig. (2). Schematic representation of the route for the preparation of (**a**) CPDB-coated SiO_2 and (**b**) sensory mechanism of the coated MIP. (Reproduced under terms of the CC-BY license [18]).

Atmospheric Water Collection

The concept of atmospheric water harvest finds its roots in a unique adaptation of Stenocara, a desert beetle, which collects water from the fog by a tailor-made covering in its body [23]. Stenocara extracts water from the air using its tailor-made carapace by inclining the body towards the direction of propelling winds. Using this same strategy to mimic the natural approach, different research groups have put forward various materials such as polymeric hydrogel, polymer brushes and metal-organic frameworks/polymer composites [6, 24, 25]. Fig. (**3**) displays the design strategy of these MOF/polymer composites [24]. For instance, Zhang *et al.* reported a porous composite hydrogel consisting of carbon fiber and polyacrylamide-co-poly(N-isopropyl acrylamide) hydrogel which was utilized for solar seawater evaporation without additional inefficient modification [26]. This composite exhibits rapid photothermal heating with the ced water evaporation enthalpy and continuous water supply. Consistently, Yang *et al.* reported a "grafting form" approach where a sponge-like cotton fabric-founded brush displayed exceptional ability to collect and release water from a humid atmosphere [27].

Fig. (3). Design and fabrication of hydrogel materials with tailored properties *via* considerations of the **a**) different chemical functionalities, **b**) network and **c**) pore structures, **d**) surface topography, and **e**) the additives to meet the multifaceted requirements for water treatment applications. (Reproduced under terms of the CC-BY license [24])

Self-Healing

The term self-healing is the realm of smart polymers having the ability to reform structural damages by themselves, thereby maintaining their original functionality [28 - 31]. In the last decade, self-healing materials have received significant attention from the scientific community due to their properties such as low damage probability, low maintenance cost, and improved safety, which prolongs the lifetime of the material for applications in challenging terrains [1].

The self-healing materials work either with the help of extrinsic components or with the help of intrinsic bond rearrangement [32]. Huang and his coworkers prepared hybrid hydrogels using PDOPA-modified clay as the primary building block (Fig. **4a**) [33]. Their results showed that the hydrogel acquires self-healing properties by reforming the damaged catechol−Fe^{3+} complex. This hydrogel can effectively remove Rhodamine 6G (Rh6G) pollutants from the waterbody by hydrogen bonding and π−π stacking interactions between clay and Rh6G. Li *et al.* developed a few self-healing elastomers by copolymerization method. They used the reversible addition− fragmentation chain-transfer (RAFT) polymerization method to prepare the functionalized Poly(BCOEr-UPy) polymers (Fig. **4b**) [34]. Tuning the fabrication of two monomers enabled to get rigid control over its mechanical properties where the tensile strength ranges from 0.04 to 6.3 MPa and tensile strain up to 3,000%. They demonstrated the additive manufacturing process where the designed self-healing elastomer was used to prepare a butterfly-shaped product on the surface in a commercially available 3D printer (Fig. **4c**). The sample displayed effective self-healing character within 5 minutes at room temperature. This study sheds light on the self-healing capability of custom-designed 3D objects.

Antifouling

Fouling, *i.e.*, unwanted adhesion, is a complex and undesirable process where, material from the environment, such as macromolecules, microorganisms, or suspended particles, adhere reversibly or irreversibly to a surface [35]. Antifouling is an additional environmental application where smart polymer materials find extensive applications. Among all the commonly used polymers for anti-fouling applications, poly(N-isopropylacrylamide) (PNIPAAM) is one of the best suited candidates [36 - 38]. Ista *et al.* in their work described the anti-fouling applications of PNIPAAM (stimuli-responsive wettability) [39]. They employed PNIPAAM-coated glass slides after 18 h of incubation with H. Marina bacteria. This PNIPAAM-coated surface effectively removed almost all the fouling bacterial cells. Apart from PNIPAAM, Kuroki *et al.* explored the activity of poly(ethylene oxide) PEO as an antifouling agent [40]. They proposed a 3D grafted polymer structure, where polymeric chains are grafted in the inner and outer surfaces of the host material. It potentially retains the antifouling effect, which is not affected by degradation and detachment of a significant fraction of the grafted polymers. Dekanovsky *et al.* explored the antifouling properties of a highly stretchable silicon polymer, gradient doped with polypyrrole [41]. This material displayed an electrically triggered release of the incorporated antimicrobial agent, leading to the killing of bacteria.

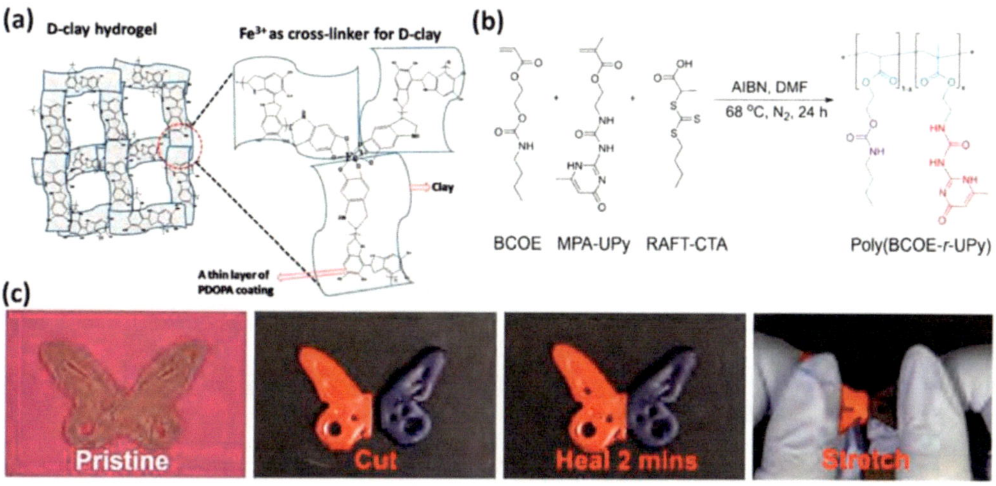

Fig. (4). (**a**) Schematic diagram of D-Clay Hydrogel formation [33]; (**b**) Preparation of self-healable random copolymer Poly(BCOE-r-UPy)s, and (**c**) Healing process of 3D-printed Poly(BCOE-r-UPy5) in different conditions [34].

CONCLUSION AND FUTURE PERSPECTIVES

From the examples discussed above, it is clear that besides the use of smart materials for solving environmental problems, they also provide ample opportunities to explore their potential in advanced device/material manufacturing. It will absolutely be able to gather sufficient attention from researchers around the world for the probable aspects of smart materials towards different applications such as catalysis, sensing, water harvesting, antifouling, and many more. Extensive research in this area will open the door for many upcoming advancements and a key role in some of the pioneering research.

REFERENCES

[1] Wang, P. *Smart Materials for Advanced Environmental Applications*; Wang, P., Ed.; Royal Society of Chemistry: Cambridge, **2013**.

[2] Ganesh, V.A.; Baji, A.; Ramakrishna, S. Smart functional polymers – a new route towards creating a sustainable environment. *RSC Advances,* **2014**, *4*(95), 53352-53364. [http://dx.doi.org/10.1039/C4RA10631H]

[3] Brighenti, R.; Li, Y.; Vernerey, F.J. Smart polymers for advanced applications: A mechanical perspective review. *Front. Mater.,* **2020**, *7*, 196. [http://dx.doi.org/10.3389/fmats.2020.00196]

[4] Mabape, K.I.S.; Donga, C.; Mishra, S.B.; Mishra, A.K. *Smart Polymer Catalysts and Their Applications*; Elsevier Inc., **2019**. [http://dx.doi.org/10.1016/B978-0-12-811840-5.00004-6]

[5] Pawar, D.; Rao, C.N.; Cao, P. Smart polymer-based chemical sensors.*Actuators*; Wiley, **2020**, pp. 75-138. [http://dx.doi.org/10.1002/9781119662693.ch5]

[6] Ritsema van Eck, G.C.; Chiappisi, L.; de Beer, S. Fundamentals and applications of polymer brushes in air. *ACS Appl. Polym. Mater.,* **2022**, *4*(5), 3062-3087. [http://dx.doi.org/10.1021/acsapm.1c01615] [PMID: 35601464]

[7] Stuart, M.A.C.; Huck, W.T.S.; Genzer, J.; Müller, M.; Ober, C.; Stamm, M.; Sukhorukov, G.B.; Szleifer, I.; Tsukruk, V.V.; Urban, M.; Winnik, F.; Zauscher, S.; Luzinov, I.; Minko, S. Emerging applications of stimuli-responsive polymer materials. *Nat. Mater.,* **2010**, *9*(2), 101-113. [http://dx.doi.org/10.1038/nmat2614] [PMID: 20094081]

[8] Gray, H.N.; Bergbreiter, D.E. Applications of polymeric smart materials to environmental problems. *Environ. Health Perspect.,* **1997**, *105*(Suppl 1) Suppl. 1, 55-63. [http://dx.doi.org/10.1289/ehp.97105s155] [PMID: 9114277]

[9] Bhawani, S.A.; Khan, A.; Jawaid, M., Eds. *Smart Polymer Nanocomposites: Biomedical and Environmental Applications*; Woodhead Publisher: Elsevier, **2021**.

[10] El-Ragehy, N.A.; El-Kosasy, A.M.; Abbas, S.S.; El-Khateeb, S.Z. Polymeric membrane electrodes for selective determination of the central nervous system acting drugs fluphenazine hydrochloride and nortriptyline hydrochloride. *Anal. Chim. Acta,* **2000**, *418*(1), 93-100. [http://dx.doi.org/10.1016/S0003-2670(00)00941-7]

[11] Liu, Z.H.; Wen, M.L.; Yao, Y.; Xiong, J. Plastic pethidine hydrochloride membrane sensor and its pharmaceutical applications. *Sens. Actuators B Chem.,* **2001**, *72*(3), 219-223. [http://dx.doi.org/10.1016/S0925-4005(00)00665-1]

[12] Xiao, C.; Chen, S.; Zhang, L.; Zhou, S.; Wu, W. One-pot synthesis of responsive catalytic Au@PVP hybrid nanogels. *Chem. Commun. (Camb.),* **2012**, *48*(96), 11751-11753. [http://dx.doi.org/10.1039/c2cc36002k] [PMID: 23108039]

[13] Arif, M. Catalytic degradation of azo dyes by bimetallic nanoparticles loaded in smart polymer microgels. *RSC Advances,* **2023**, *13*(5), 3008-3019. [http://dx.doi.org/10.1039/D2RA07932A] [PMID: 36756456]

[14] Qi, J.; Lv, W.; Zhang, G.; Li, Y.; Zhang, G.; Zhang, F.; Fan, X. A graphene-based smart catalytic system with superior catalytic performances and temperature responsive catalytic behaviors. *Nanoscale,* **2013**, *5*(14), 6275-6279. [http://dx.doi.org/10.1039/c3nr00395g] [PMID: 23740038]

[15] Reglero Ruiz, J.; Sanjuán, A.; Vallejos, S.; García, F.; García, J. Smart polymers in micro and nano sensory devices. *Chemosensors (Basel),* **2018**, *6*(2), 12. [http://dx.doi.org/10.3390/chemosensors6020012]

[16] Peeters, M.; Kobben, S.; Jiménez-Monroy, K.L.; Modesto, L.; Kraus, M.; Vandenryt, T.; Gaulke, A.; van Grinsven, B.; Ingebrandt, S.; Junkers, T.; Wagner, P. Thermal detection of histamine with a graphene oxide based molecularly imprinted polymer platform prepared by reversible addition–fragmentation chain transfer polymerization. *Sens. Actuators B Chem.,* **2014**, *203*, 527-535. [http://dx.doi.org/10.1016/j.snb.2014.07.013]

[17] Wan, W.; Biyikal, M.; Wagner, R.; Sellergren, B.; Rurack, K. Fluorescent sensory microparticles that "light-up" consisting of a silica core and a molecularly imprinted polymer (MIP) shell. *Angew. Chem.*

Int. Ed., **2013**, *52*(27), 7023-7027.
[http://dx.doi.org/10.1002/anie.201300322] [PMID: 23716378]

[18] Valderrey, V.; Gawlitza, K.; Rurack, K. Thiourea- and amino-substituted benzoxadiazole dyes with large stokes shifts as red-emitting probe monomers for imprinted polymer layers targeting carboxylate-containing antibiotics. *Chemistry,* **2022**, *28*(21), e202104525.
[http://dx.doi.org/10.1002/chem.202104525] [PMID: 35224792]

[19] Shammas, N.K., Ed. *Membrane and Desalination Technologies; Wang, L. K., Chen, J. P., Hung, Y.-T*; Humana Press: Totowa, NJ, **2011**.

[20] Zhang, W.; Shi, Z.; Zhang, F.; Liu, X.; Jin, J.; Jiang, L. Superhydrophobic and superoleophilic PVDF membranes for effective separation of water-in-oil emulsions with high flux. *Adv. Mater.,* **2013**, *25*(14), 2071-2076.
[http://dx.doi.org/10.1002/adma.201204520] [PMID: 23418068]

[21] Gao, X.; Xu, L.P.; Xue, Z.; Feng, L.; Peng, J.; Wen, Y.; Wang, S.; Zhang, X. Dual-scaled porous nitrocellulose membranes with underwater superoleophobicity for highly efficient oil/water separation. *Adv. Mater.,* **2014**, *26*(11), 1771-1775.
[http://dx.doi.org/10.1002/adma.201304487] [PMID: 24347397]

[22] Jiang, G.; Zhang, C.; Xie, S.; Wang, X.; Li, W.; Cai, J.; Lu, F.; Han, Y.; Ye, X.; Xue, L. Facile fabrication of electrospun nanofibrous aerogels for efficient oil absorption and emulsified oil–water separation. *ACS Omega,* **2022**, *7*(8), 6674-6681.
[http://dx.doi.org/10.1021/acsomega.1c06080] [PMID: 35252662]

[23] Parker, A.R.; Lawrence, C.R. Water capture by a desert beetle. *Nature,* **2001**, *414*(6859), 33-34.
[http://dx.doi.org/10.1038/35102108] [PMID: 11689930]

[24] Loo, S.L.; Vásquez, L.; Athanassiou, A.; Fragouli, D. Polymeric hydrogels—a promising platform in enhancing water security for a sustainable future. *Adv. Mater. Interfaces,* **2021**, *8*(24), 2100580.
[http://dx.doi.org/10.1002/admi.202100580]

[25] Matemb Ma Ntep, T.J.; Wahiduzzaman, M.; Laurenz, E.; Cornu, I.; Mouchaham, G.; Dovgaliuk, I.; Nandi, S.; Knop, K.; Jansen, C.; Nouar, F. When polymorphism in metal–organic frameworks enables water sorption profile tunability for enhancing heat allocation and water harvesting performance. *Adv. Mater.,* **2023**, 2211302.
[PMID: 36897806]

[26] Zhang, P.; Ye, J.; Li, G.; Si, Y.; Qian, C.; Fu, Y. Rapid fabrication of porous composite hydrogels for efficient solar vapor generation. *ACS Appl. Polym. Mater.,* **2023**, *5*(8), 6633-6642.
[http://dx.doi.org/10.1021/acsapm.3c01202]

[27] Yang, H.; Zhu, H.; Hendrix, M.M.R.M.; Lousberg, N.J.H.G.M.; de With, G.; Esteves, A.C.C.; Xin, J.H. Temperature-triggered collection and release of water from fogs by a sponge-like cotton fabric. *Adv. Mater.,* **2013**, *25*(8), 1150-1154, 1149.
[http://dx.doi.org/10.1002/adma.201204278] [PMID: 23303561]

[28] Terryn, S.; Langenbach, J.; Roels, E.; Brancart, J.; Bakkali-Hassani, C.; Poutrel, Q.A.; Georgopoulou, A.; George Thuruthel, T.; Safaei, A.; Ferrentino, P.; Sebastian, T.; Norvez, S.; Iida, F.; Bosman, A.W.; Tournilhac, F.; Clemens, F.; Van Assche, G.; Vanderborght, B. A review on self-healing polymers for soft robotics. *Mater. Today,* **2021**, *47*, 187-205.
[http://dx.doi.org/10.1016/j.mattod.2021.01.009]

[29] Irzhak, V.I.; Uflyand, i.e.; Dzhardimalieva, G.I. Self-healing of polymers and polymer composites. *Polymers (Basel),* **2022**, *14*(24), 5404.
[http://dx.doi.org/10.3390/polym14245404] [PMID: 36559772]

[30] Dzhardimalieva, G.I.; Yadav, B.C.; Kudaibergenov, S.E.; Uflyand, i.e. Basic approaches to the design of intrinsic self-healing polymers for triboelectric nanogenerators. *Polymers (Basel),* **2020**, *12*(11), 2594.
[http://dx.doi.org/10.3390/polym12112594] [PMID: 33158271]

[31] Li, B.; Cao, P.F.; Saito, T.; Sokolov, A.P. Intrinsically self-healing polymers: from mechanistic insight to current challenges. *Chem. Rev.,* **2023**, *123*(2), 701-735.
[http://dx.doi.org/10.1021/acs.chemrev.2c00575] [PMID: 36577085]

[32] Peng, Y.; Gu, S.; Wu, Q.; Xie, Z.; Wu, J. High-performance self-healing polymers. *Acc. Mater. Res.,* **2023**, *4*(4), 323-333.
[http://dx.doi.org/10.1021/accountsmr.2c00174]

[33] Huang, S.; Yang, L.; Liu, M.; Phua, S.L.; Yee, W.A.; Liu, W.; Zhou, R.; Lu, X. Complexes of polydopamine-modified clay and ferric ions as the framework for pollutant-absorbing supramolecular hydrogels. *Langmuir,* **2013**, *29*(4), 1238-1244.
[http://dx.doi.org/10.1021/la303855t] [PMID: 23289718]

[34] Li, B.; Ge, S.; Zhao, X.; Chen, Q.; Tian, J.; Hun, D.; Sokolov, A.P.; Saito, T.; Cao, P.F. Well-tunable, 3D-printable, and fast autonomous self-healing elastomers. *Supramolecular Materials,* **2023**, *2*, 100042.
[http://dx.doi.org/10.1016/j.supmat.2023.100042]

[35] Maan, A.M.C.; Hofman, A.H.; de Vos, W.M.; Kamperman, M. Recent developments and practical feasibility of polymer-based antifouling coatings. *Adv. Funct. Mater.,* **2020**, *30*(32), 2000936.
[http://dx.doi.org/10.1002/adfm.202000936]

[36] Ista, L.K.; Mendez, S.; Pérez-Luna, V.H.; López, G.P. Synthesis of poly(*N*- isopropylacrylamide) on initiator-modified self-assembled monolayers. *Langmuir,* **2001**, *17*(9), 2552-2555.
[http://dx.doi.org/10.1021/la001257d]

[37] Galaev, I.; Mattiasson, B. 'Smart' polymers and what they could do in biotechnology and medicine. *Trends Biotechnol.,* **1999**, *17*(8), 335-340.
[http://dx.doi.org/10.1016/S0167-7799(99)01345-1] [PMID: 10407406]

[38] Xu, X.; Liu, Y.; Fu, W.; Yao, M.; Ding, Z.; Xuan, J.; Li, D.; Wang, S.; Xia, Y.; Cao, M. Poly(N-isopropylacrylamide)-based thermoresponsive composite hydrogels for biomedical applications. *Polymers (Basel),* **2020**, *12*(3), 580.
[http://dx.doi.org/10.3390/polym12030580] [PMID: 32150904]

[39] Ista, L.K.; Lopez, G.P. Lower critical solubility temperature materials as biofouling release agents. *J. Ind. Microbiol. Biotechnol.,* **1998**, *20*, 121-125.
[http://dx.doi.org/10.1038/sj.jim.2900490]

[40] Kuroki, H.; Tokarev, I.; Nykypanchuk, D.; Zhulina, E.; Minko, S. Stimuli-responsive materials with self-healing antifouling surface *via* 3D polymer grafting. *Adv. Funct. Mater.,* **2013**, *23*(36), 4593-4600.
[http://dx.doi.org/10.1002/adfm.201300363]

[41] Děkanovský, L.; Elashnikov, R.; Kubiková, M.; Vokatá, B.; Švorčík, V.; Lyutakov, O. Dual-action flexible antimicrobial material: switchable self-cleaning, antifouling, and smart drug release. *Adv. Funct. Mater.,* **2019**, *29*(31), 1901880.
[http://dx.doi.org/10.1002/adfm.201901880]

CHAPTER 5

Application of Smart Polymeric Composites/Nanocomposites in Drug Delivery

Moon Mandal[1,*] and **Zakaria Halim**[2]

[1] *Department of Chemistry, Pandit Deendayal Upadhyaya Adarsha Mahavidyalaya, Dalgaon, Darrang-784116, Assam, India*

[2] *Department of Chemical Engineering, Golaghat Engineering College, Bogorijeng, Golaghat-785621, Assam, India*

Abstract: Over the course of several decades, an interest in stimuli-responsive polymers has grown significantly, leading to extensive research in the development of environmentally responsive macromolecules, which can be exhibited as smart novel polymers. They have emerged as a promising platform for the advancement of drug delivery systems, enhancing the efficiency of therapeutic interventions, decreasing dosage frequency, minimizing possible adverse reactions and enhancing patient receptivity. In particular, these materials are responsive to very subtle changes in their environment. The environmental triggers for such transformations can include temperature change, shift in pH, variations of ionic strength, alterations in magnetic or electric fields, exposure to light, *etc*. They are said to be stimuli-responsive or intelligent polymers. Smart polymeric materials provide a versatile platform for drug delivery, enabling the controlled release of drugs in a stable and bioactive form. Additionally, the incorporation of nanocomposites contributes to increased drug bioavailability, enables targeted delivery through surface modification, and facilitates sustained release profiles. This chapter mainly provides detailed progress of smart polymeric composites or nanocomposites in the advancement of drug delivery applications. It also thoroughly covers several characteristics of these polymers categorized into different groups based on the stimulus type. The chapter aims to shed light on the evolving landscape of stimuli-responsive polymers and their pivotal role in shaping the future of drug delivery systems.

Keywords: Drug delivery, Intelligent polymers, Nanocomposites, Stimuli-responsive, Smart polymers.

INTRODUCTION

The polymers in today's world play a crucial role, and it is difficult to imagine life without them. They are a group of advanced materials with uses in various areas

* **Corresponding author Moon Mandal:** Department of Chemistry, Pandit Deendayal Upadhyaya Adarsha Mahavidyalaya, Dalgaon, Darrang-784116, Assam, India; E-mail: mnmandal8@gmail.com

Partha Pratim Saikia & Bhaskar Jyoti Saikia (Eds.)

of science, technology and industry – from fundamental applications to biopolymers or therapeutic purposes. Polymers are available naturally and are also man-made. The study of man-made polymeric materials became very prominent during the mid-19th century. Since then, the polymer industry today has seen rapid growth and is now greater than the combined industries of steel, copper, aluminum, *etc* [1].

A polymer is a large molecule (macromolecule) made up of long chain repeating units of small molecules. These small molecules are usually connected by covalent chemical bonds and these repeating units are called monomers [2]. The awareness of various kinds of polymers and their applications among people emerged only after World War II [1]. Depending on the type of molecules bonded, they have unique properties and applications. Polymers in today's world have found applications ranging from transportation to medicine and electronics, clothes and cookware to housing, *i.e.* the impact of polymer in the present way of life is immeasurable [3]. Because of advantages such as ease of processing, better strength-to-weight ratios, durability against chemical and physical degradation, and cost-effectiveness, polymers have replaced a number of conventional materials in various applications [4]. As such, polymers serve as one of the primary contributors to the industry as well as the economy. This is evident from the fact that Europe had a production volume of 57.9 million tons of polymers in 2019, which accounts for 16% of the world's total production with Asia being the leading producer [5].

The inspiration and intelligence of nature in recent times have led to the development of the smartness of material. The polymers that integrate the combined functions of sensing, actuation, and control can be defined as "smart polymers". Amidst the era of the 4th industrial revolution, researchers have drawn a huge interest in smart polymers [6]. These polymers respond to mild alterations in chemical as well as physical states with sharp and relatively large phase or property changes. The responses to external stimuli result in macro-scale changes like contraction, expansion, twisting, elongation, or rolling of the materials [7]. As such, they are also regarded as "environmentally sensitive" polymers. These polymers have the ability to take many forms, and can be physically or chemically modified with other bioactive molecules as well. Over the past 20-25 years, smart polymers have found extensive uses in the biomedical field [8]. Smart polymers in the medical field are used for the treatment of several diseases and also in advanced medical sectors. They respond to temperature, pH, enzymes, and concentration of analytes, light, or magnetic field. Smart polymers have also found applications in bioseparation, data storage devices, and rechargeable batteries to name a few. Thus, the emerging research in the field of smart polymer has not only opened avenues for their applications in the medical field but also in the food sector, architecture, data, and energy storage [9]. This chapter mainly focuses on the significant advancements made by smart polymeric nanocomposites for drug delivery applications, as it undergoes a thorough exploration, covering diverse aspects of these stimuli-responsive materials. It also offers comprehensive insights into their evolving role and impact on drug delivery systems, by highlighting the crucial contributions of these smart materials in drug delivery technologies.

POLYMER NANOCOMPOSITES

Due to the various properties of polymers and their composites like lightweight, low cost, excellent corrosion resistance, easy processing, *etc*., polymer composites have widespread applications in all aspects of daily life and industry [10]. Polymer composites are a combination of polymers and a material in which the final product has superior properties compared to each constituent material. A polymer composite usually consists of a continuous phase (matrix) and a discontinuous phase (reinforcement) [11]. Polymer nanocomposites on the other hand are polymer composites with fillers having any one dimension less than 100 nm. They usually have very minimal amounts of well-dispersed nanofillers compared to polymer composites [12].

SMART POLYMER NANOCOMPOSITES

Smart polymer composites and nanocomposites are an advanced class of materials, which respond to environmental stimuli like light, enzyme, pH, or electric field, resulting in better performance and increased efficiency. Table **1** shows some examples of stimuli-responsive polymers and their significant results in the biomedical sector [13]. Before the application of smart polymer composites and nanocomposites, there is a need for extensive research on the capabilities of smart polymer composites/nanocomposites to revolutionize several industries [14]. The ability of smart polymer composites and nanocomposites to respond to environmental stimuli provides various advantages like greater performance, increased efficiency, and self-healing capacity. Some widely used smart polymer composites and nanocomposites are shape-memory, self-healing, and conductive polymers. The shape memory polymers have the ability to revert back to their original character as the deforming force is withdrawn. Conductive polymers can conduct electricity and are used in applications such as actuators or sensors. Self-healing polymers are those, which can repair damage like cracks or fractures caused by external factors, thereby extending their lifespan [15].

Table 1. Examples of Stimuli-Responsive Polymers.

Stimuli	Polymer	Results
pH	Poly(lactic acid)-poly(ethyleneimine)	Doxorubicin (DOX) is rapidly released when there is a change in pH from 7.4 to 5.4.
Ultrasound	Chitosan	Notable increase in release compared to situations where no stimulus is applied.
Near Infrared (NIR) light	Polycaprolactone	Drug-release in multiple steps following repeated cycles of exposure to NIR, with low off-state leakage.
Ultraviolet (UV)	mPEG-PLGA nanoparticles	Increase the sensitivity of tumour cells to Doxorubicin therapy by reversing multidrug resistance.
Electric	Poly(3-methoxydiphenylamine)/Pectin blend	Stimulation leads to an increase in mesh pore size of hydrogel, resulting in enhanced elution of drug.
Enzyme	Poly(ethylene glycol)	Cleaving peptides at specific sites.
Permanent magnet	Polyethyleneimine	External magnetic guidance enhanced the accumulation of superparamagnetic iron oxide nanoparticles in arthritic joints within a model rat.

The shape memory polymer composites/nanocomposites are those which revert back to their original character after deformation and respond to stimuli because of their shape-shifting behaviour. Usually, reinforcing agents like metal oxides, ceramics, metals, silica, carbons, crystalline polymers, elastomers, and more are used in shape memory polymer composites. The phase transformations are influenced by various ranges of temperature [16]. In general, shape memory polymers are characterized by permanent net points, which form a stable network of polymer, and by molecular switches that are reversible in nature. The net-points memorize the initial shape and the molecular switches are responsible for establishing temporary shape. The shape memory polymers when subjected to physical blending, chemical cross-linking, or *in-situ* polymerization with nanofillers can enhance their mechanical properties and shape recovery stress. Shape memory nanocomposites can be used in various applications like electrochemical devices, biomimetic devices, smart textiles, smart windows, actuators and sensors, active sound-absorbing materials, deployable space structures, morphing aircraft, artificial muscle, civil and architecture applications [6]. Electrically conductive polymer composites (ECPCs) have emerged as promising candidates for smart flexible strain sensors, responding to alterations in electrical resistance under various external strains like bending, compression, stretching, and twisting. Due to the polymer matrix's excellent flexibility and stretching capacity, these sensors can be knitted, skin-attached, and can be worn. By using suitable processing technology, ECPCs are fabricated by blending conductive fillers with an insulated polymer matrix. ECPCs show significant potential for the preparation of wearable electronic devices and are seen as ideal alternatives to conventional strain sensors [17].

Self-healing operates at a microscale level, using bottom-up methods to repair damages and seal cracks while ensuring the materials' structural integrity. Self-healing polymer nanocomposites can autonomously repair damages caused by mechanical friction [18]. Intrinsic self-healing and extrinsic self-healing are the two ways of healing strategies. In the first strategy, the polymer can heal damages such as cracks using its thermally reversible bonds, which induce a healing effect when subjected to external heating. However, in the second strategy, the polymer is infused with a healing agent like microcapsules, microvascular networks, hollow fibers, *etc* [6]. Self-healing anti-corrosion coatings, self-healing mechanisms in concrete, and repairing surface cracks in ceramics are some of the applications of polymer composites/nanocomposites with self-healing abilities. Wearable health monitoring systems have gained significant interest recently due to their portable nature and broad medical applications. The sensors used in them possess excellent flexibility and stretchability and provide health investigative systems of continuous monitoring of physiological signals from the human body without causing any inconvenience or intrusion. Signals like the rate of the heart, the motion of the body, breathing, temperature of the skin, and metabolism, make them closely associated with individual health conditions. Due to the improvement of materials' shelf life and reduction in the long-term system cost during application and use, self-healing polymer composites/nanocomposites have sparked global research interest [18]. Overall, owing to their multidimensional properties, nanocomposites are promising candidates in the context of drug delivery applications.

TYPES OF SMART MATERIALS

Shape Memory Polymer Nanocomposites

Shape memory polymers (SMPs) are dynamic polymers that memorize the initial shape, hold a deformed shape temporarily, and revert to their initial shape upon encounter with a stimulus. The most important stimuli for SMP are heat and magnetic field, other triggers include, light, water, solvent, pH, temperature, and electricity. SMPs also have a few unusual properties like huge recoverable stroke, pseudo-elasticity, and elevated damping ability. They can react to temperature, mechanical, magnetized, or electrical stimulus to regulate specific technical factors like strain, configuration, rigidity, natural frequency, position, resistance, damping, and different dynamic as well as static features of the system. Although SMPs possess various advantages, a major drawback they face is due to their inadequate mechanical properties. Actuation usually relies on thermal management, and minimal recovery stress because of their inferior rubbery modulus. Some other drawbacks of SMPs include limited thermal conductivity, inactivity to light, electrical or magnetic stimulus along with steady response, and minimal time of recovery while actuation that usually limits significant application of these materials. Shape Memory Polymer Nanocomposites (SMPNCs) improve the shape recovery stress as well as mechanical characteristics of SMPs, while also decreasing the shape recovery time by enhancing thermal conductivity by adding nano-materials. They are useful in adjusting the range of switching temperature and tensile properties of the SMPs. Recently, various nanosized fillers like carbon nanotubes (CNTs), quantum dots, metal oxide nanoparticles, cellulose nano-crystals, nano-cellulose gel, *etc*. were incorporated to increase the properties of SMPs. These smart materials are extensively used in smart textiles, medical sectors, sensors, autonomous structures in spacecraft, *etc*. They are also used for various biomedical applications like wound dressing, bone tissue engineering, drug delivery, bone defect repair, *etc* [19].

Self-healing Polymer Nanocomposites

Self-healing polymers (SHPs) are types of smart materials, that have the ability to heal, and undergo *in-situ* damage restoration and property renewal effortlessly. This helps in extending the service life, increasing product reliability, reducing the frequency of scheduled inspections, and lowering the maintenance cost of the final material. There are two methods of healing strategies *i.e*., intrinsic self-healing and extrinsic self-healing. In the intrinsic self-healing method, the polymer itself heals the damages or cracks, at which the inherent thermally reversible bonds of the substance develop a recovering effect on external heating. In the extrinsic self-healing method, self-healing occurs using pre-embedded healing agents like microcapsules, microvascular networks, hollow fibers, *etc*. The self-healing efficacy serves as one of the pivotal performance metrics for SHP nanocomposites that are regulated by various aspects like the polymer matrices, nano-filler choice, selection of self-healing mechanism, and the environment. Microbial diseases are a major threat to human health. In this regard, Zhao *et al*. prepared polysiloxane nanocomposites that have self-healing features, utilizing furan-modified graphene as reinforcement. The healable polysiloxane nanocomposite retrieved nearly 90 to 98% of its tensile strength and elongation at break, respectively. In another study, self-healing nanocomposite showed significant antimicrobial activity *versus Staphylococcus aureus*, *Escherichia coli*, and *Candida albicansas,* as it contained sulfur-loaded substances and polysulfanes that were regarded as antimicrobial agents. These nanocomposites have gained popularity mostly in the medical areas, owing to their capacity as carriers for anticancer drugs. Also, hydrogels exhibited promising materials for several environmental and biological applications on account of DNA biocompatibility [20, 21].

pH-sensitive Polymer Nanocomposites

pH-sensitive polymer nanocomposites are a versatile class of materials with various applications in several areas, due to their unique properties derived from the polymer matrix and nanofiller components. They have shown significant success owing to their ability to respond to changes in pH, making them suitable for applications like site-specific delivery of drugs. In drug delivery applications, they offer precise control over drug release kinetics, enabling targeted drug delivery, and enhanced therapeutic efficacy. Different types of pH-sensitive polymers are commonly used in nanocomposite development, including hydrogels, polyelectrolytes, *etc*. For example, Barkhordari *et al*., prepared nanocomposite beads that were pH-sensitive by combining layered double hydroxides (LDH) and carboxymethyl cellulose (CMC). A model drug, Ibuprofen (IBU) was intercalated amidst LDH *via* co-precipitation methodology. To prove the effectiveness of this novel type of nanocomposite beads, an *in vitro* study of drug delivery under environments simulating the gastrointestinal tract was conducted. The drug liberation study showed an improved defense in contrast to stomach pH and a regulated release in the intestinal tract on ditions for the prepared beads [22]. In another report, a nanocomposite was fabricated by adding various concentrations of chitosan-modified carbon nanotubes into acrylamide-c--acrylic acid hydrogels, which showed improvement in the mechanical properties and showed robust pH-response [23].

Temperature-responsive Polymer Nanocomposites

Temperature-responsive polymer nanocomposites are smart materials, which have the ability to react to temperature variances. They are useful in various fields, including biomedical, sensors, smart coatings, *etc*. They are the most studied class of environmentally sensitive polymers with respect to cost-effectiveness, biocompatibility, thermal constancy, and greater flexibility. One of the major features of temperature-responsive materials is their ability to undergo reversible phase transitions in response to temperature changes. Based on their response to temperature change, these polymers are categorized into lower critical solution temperature (LCST) and upper critical solution temperature (UCST). Polymers that become insoluble above a critical temperature are known as LCST. Some examples include, poly(*N*- vinyl caprolactam), poly(*N, N* -dimethyl acrylamide), poly oxazoline, poly[2-(dimethylamino)ethyl methacrylate], block copolymers of polyethylene oxide, and poly(*N* -isopropyl acrylamide).

Polymers that precipitate and undergo phase change below a critical temperature are UCST. Poly(acrylamide) and poly(acrylic acid) copolymers are some examples. In a study, a thermoresponsive polymer clay nanocomposite with better temperature response derived from organically modified clay-poly(*N*-isopropyl acrylamide) was fabricated by Liang *et al.* This material showed huge thermal volumetric transformation and rapid transition kinetics [24]. Similarly, in another study, temperature-sensitive, antibacterial poly(di(ethylene glycol)methyl ether methacrylate) – POEGMA188 derived nanocomposite coatings connected to a glass surface were prepared by atom transfer radical polymerization. The material was reported to be non-toxic to normal cells and had strong temperature-dependent antibacterial activity [25].

Light-responsive Polymer Nanocomposites

Light-responsive polymer nanocomposites are materials that exhibit changes in their properties or behavior upon exposure to light. Light *i.e.*, infrared, UV radiation, or simply sunlight is considered to be a highly convincing external stimulus, as it is usually sent without direct interaction, thereby exhibiting one of the rare remote-control prompting features obtainable. Photo-responsive substances, like other potential technologies, have been influenced by nature. Nature has developed an intricate biological system that ventures light as an external source of power and data. Photo-induced cis-trans isomerization of the retinal molecule activates several situations, like changes in the configuration of the opsin protein that leads to the generation of a neural signal and then sensation of the light. Photo-responsive polymers are formed by mimicking natural structures by launching light-sensitive components into the polymer. According to reports, azobenzene, stilbene, and psicopyranose are some of the examples of light-sensitive compounds. The photo-sensitivity of the functional groups of these molecules relies on the presence of two isomers that are interconvertible. Upon exposure to light, usually in the UV region, the molecules experience some rearrangement in their conformation. For azobenzene and stilbene, the change is showcased through shifts in the symmetry of the molecule. It changes from trans (E) conformation (thermally stable) to cis (Z) conformation (less favorable). In the case of spiropyrans, the irradiation undergoes a ring-opening reaction, which leads to the formation of an isomeric merocyanine state [26]. Chen *et al.*, conducted a study wherein they fabricated shape memory polymer nanocomposites with rapid light-sensitivity and self-healing capabilities. This work has received prompt focus owing to its numerous promising uses. They prepared a blend of poly(ε-caprolactone) (PCL) and thermoplastic polyurethane (TPU) as the shape memory matrix *via* a solution blending process. Subsequently, polydopamine (PDA) nanospheres as photo-thermal fillers were inserted in the PCL/TPU blend that helps the system to give outstanding photo-response shape memory results. The shape recovery ratio of the system was reported to be 100%. Moreover, the cracks on the surface were found to be repaired in 150 seconds in the presence of light [27]. Likewise, a nano platform comprising hybrid mesoporous silica (MS)--covered carbon nanotubes (CNTs) was formed to facilitate phototherapy along with drug release mediated by near-infrared light. It was reported that the nanocomposites showed promising results as a new component of implantable antitumor scaffolds, capable of timely and localized responses to external stimuli, thereby enhancing disease management [28].

Magnetic Field-Responsive Polymer Nanocomposites

Magnetic field-responsive polymer nanocomposites are attracting greater focus due to their breakthrough uses in the biomedical fields. They could be readily altered and configured; also they may be controlled externally by the system. Moreover, they can function in various mediums like vacuum, air, conducting, or non-conducting liquids, provided the actuation atmosphere is transparent magnetically. A promising approach in the development of advanced polymer nanocomposites involves the incorporation of functional nanoparticles, such as magnetic nanoparticles (MNPs) that exhibit a response to the external magnetic field. The filler magnetic sensitivity in combination with the capacity to design polymers for specific performance offers new opportunities to develop a range of multifunctional materials capable of adapting to changing external conditions. For instance, in the case of elastomers exhibiting comparatively low modulus of elasticity (below 200kPa, when magnetic interactions can effectively compete with the elastic characteristics of the matrix), variation in the magnetic field strength can influence the shape of the material. The rearrangement of MNPs induced by the field can also generate localized anisotropy in the distribution of the filler. Therefore, it should lead to the anisotropy of the mechanical properties during uniaxial or biaxial mechanical deformation [29]. In a study, a novel magnetic responsive polymer nanocomposite (MRPNCs) was developed with *in-situ* adjustable magnetic anisotropy. These nanocomposites can undergo self-organization at a lower magnetic field by regulating the self-organization temperature within the melting point range between isotactic polypropylene (iPP) and poly(ε-caprolactone) (PCL). Magnetic ferro ferric oxide (Fe_3O_4) nanoparticles were selectively dispersed in PCL, which has a lower melting temperature. The resulting iPP/PCL/Fe_3O_4 nanoparticles exhibit magnetic anisotropy, allowing tunable rotation behavior and mechanical enhancement. The iPP/PCL/Fe_3O_4 nanocomposites with magnetic anisotropy show significantly improved properties compared to samples with random dispersion of magnetic nanoparticles. Nanocomposites containing 10 wt% Fe_3O_4 showed approximately 3.2 times higher deflection force, 3.2 times more tensile strength and 2.7 times greater Young's modulus [30]. Similarly, another report focused on reprogrammable, magnetically controlled polymeric nanocomposite actuators. In this study, the actuation performance and kinetics were found to be controlled by precise adjustments in both the applied magnetic field strength and exposure time [31].

Electric Field Responsive Polymer Nanocomposites

Electro-active polymers (EAPs) are a group of materials that may be fabricated by introducing conductive phases such as carbon nanoparticles (CNPs), carbon nanofiber, or carbon nanotubes into the polymer matrix, which could effectively improve both the mechanical and electrical properties. The EAPs are usually defined by their responsive behavior to the electric voltage or electric field. A synergistic effect was proposed for SMPs consisting of shape memory and electro-responsive properties to enhance better responsive performances, achieving higher actuation stain (up to 400%) and controllable recovery time. Additionally, the SMP-based EAPs with favorable dielectric properties hold promise for applications in bio-robots, mini-pumps, and orthotic/prosthetics due to their controllable electrical triggering behavior [32]. Chandran and Sandhyarani introduced a novel drug delivery system using chitosan-gold nanocomposite (CGNC) that responds to both electric fields and pH. This system is fabricated to deliver the anticancer drug5-fluorouracil (5-FU) to specific sites in a controlled way. At a higher level, CGNC encompasses the drug within its three-dimensional network, releasing them fairly at minimum pH levels because of their reversible gel-sol transition features. Sulphorhodamine B cytotoxicity test results demonstrated greater cytotoxicity of CGNC-FU conjugate system against cervical cancer cells (SiHa) compared to 5-FU alone. The release of the drug from the nanocomposite was regulated *via* a fragile external electric field using direct current. The encapsulation efficacy of the nanocomposite was reported to be 36%, with approximately 63% of the drug being discharged by the application of an electric field. The biocompatibility of nanocomposite was verified through SiHa cell proliferation on a CGNC film electrodeposited onto an indium tin oxide (ITO) covered glass plate. Moreover, SiHa cells cultured on a CGNC-FU conjugate-modified ITO plate exhibited reduced viability. The application of a further 1.5V electric field to release 5-FU resulted in total cell demise [33].

PROPERTIES OF SMART POLYMER NANOCOMPOSITES

Responsive to Stimuli

A responsive nanocomposite material is usually fabricated to react spontaneously by altering the structure of soft material, utilizing a combination of organic and inorganic components with stimuli-responsive characteristics. These smart nano-composite polymers show sensitivity towards pH, heat, magnetism, and electric field, *etc.*, for example, heat-responsive nanocomposites are mostly applied as smart materials because temperature is the simplest external stimulus known. Hydrogels are prominent examples of thermoresponsive polymers, displaying a reversible phase transformation of polymer chains (from hydrolyzed to precipitated form). It is also accompanied by a notable shift in volume in context to temperature changes. The NCs that respond to magnetic fields are influenced by static or alternating magnetic fields. An advantage of such NCs is that these magnetic fields possess comparatively huge permeation depth that provides a contactless way of stimulation. These materials may be utilized as magnetically separable substances in the targeted delivery of drugs and purification. The use of electric currents plays a fascinating role in smart materials as they can induce a change in the materials' shape. Carbon-derived nanomaterials, such as graphene, graphene oxide, or carbon nanotubes are noteworthy nanofillers due to their sensitivity to electrical current, exhibiting excellent mechanical and thermal properties with improvement in the electronic properties as well. In a study, a polymer blend comprising 2-acrylamide-2-methyl propane sulfonic acid (AMPS) and acrylamide (AAm), reinforced with reduced graphene oxide (rGO) as filler was prepared. They employed an *in-situ* polymerization technique to incorporate nanoplatelets of GO into the polymer matrix, followed by reduction of GO using hydrazine to gain rGO/poly(AMPS-co-AAm) nanocomposite. This nanocomposite exhibited electrically controllable swelling/deswelling behavior. Moreover, compared to a pure polymer matrix, the nanocomposite showed enhancements in both mechanical and electrical response [34, 35].

Tunable Mechanical Properties

One of the important properties of smart polymer nanocomposites is their unique mechanical properties that can be tuned or controlled. By incorporating nanoparticles or nanofillers into the polymer matrix, properties like stiffness, strength, elasticity, and toughness can be modified. There are reports on the development of an injectable, self-repaired, highly pH-sensitive nanocomposite hydrogel that relies on the utilization of reactions (Schiff base) among aldehyde-modified polymers and amine-functionalized silica nanoparticles. This approach enables the creation of a versatile hydrogel system capable of responding to changes in pH levels. By incorporating amine-modified silica nanoparticles into the aldehyde-functionalized polymer matrix, a robust nanocomposite structure was formed, offering enhanced mechanical properties and improved sensitivity to environmental stimuli. The resulting hydrogel exhibited remarkable self-healing capabilities, allowing it to autonomously repair damage and maintain its integrity over time. Additionally, its injectable nature facilitates easy application in various biomedical and drug delivery applications, making it a promising candidate for advanced therapeutic interventions. The hydrogel exhibited rapid gelation, achieving full formation within just 10 seconds, along with injectability and remarkable self-healing capabilities. Moreover, it showed exceptional consistency in a neutral physiological state, experiencing a rapid gel-sol transformation in response to a slightly acidic atmosphere, making it ideal for regulated drug release applications. The pH sensitivity of the system was exceptionally strong with mechanical characteristics, hydrolytic breakdown, and drug delivery performance considerably changing in response to mild pH variation. Furthermore, the mechanical properties and pH-reactivity of the hydrogel could be easily adjusted by modifying the structure, offering versatility and adaptability for various biomedical and therapeutic applications [36].

Enhanced Stability and Durability

A smart polymeric material has the ability to withstand prolonged exposure to harsh environmental conditions, like mechanical stress and chemical degradation without significant deterioration in its properties. The durability is influenced by factors such as the type of polymer matrix, dispersal of nanoparticles, and interfacial interaction among the polymer and nanoparticles. Stability means the capacity of the material to maintain its properties at elevated temperatures, and chemical stability, which involves resistance to degradation from exposure to solvents, acids, bases, and other environmental factors. In a study, Wang and Guo developed a smart superhydrophobic surface by fabricating a UV light source healed fluorinated polymer layer with outstanding strength and flame retardation. The coating was formulated by single suspension having trimethylolpropane triacrylate, vinyl tri-ethoxysilane, 1H, 1H, 2H, 2H-Perfluorododecanethiol, 9,10-Dihydro-9-Oxa-10-Phosphaphenanthrene-10-Oxide, SiO_2 nano-particles, and 2-Hydroxy-2-methylpropiophenone. The formulation has been directly sprinkled onto different substrates and subsequently cured under ultraviolet lights. The resultant material exhibited great water resistance, mechanical and chemical endurance and flame resistance [37].

Improved Thermal and Electrical Conductivity

Thermal conductivity is considered to be an important property in different applications ranging from heat dissipation in electronics to thermal management in advanced materials. A smart material offers the potential to enhance thermal conductivity through the addition of thermally conductive nanofillers like carbon nanotubes, graphene or metallic nanoparticles. The dispersal of these nanofillers in the matrix of the polymer allows efficient heat transfer, thereby increasing the overall thermal conductivity of the material. Also, using stimuli-responsive polymers enables the modulation of thermal conductivity with respect to external stimuli like temperature, heat, electric field, *etc.* further expanding the range of applications in these materials. Similarly, electrical conductivity also plays a vital role in various applications like flexible electronics, sensors, *etc*. Smart materials can also increase their electrical conductivity by adding nanofillers like graphene, CNT, conducting polymers, *etc*. The dispersion of nanofillers within the polymer matrix allows efficient charge transport throughout the nanocomposite. The performance of smart materials with respect to thermal and electrical conductivity is influenced by several factors like the type of nanofillers selected, properties of the polymer matrix, processing techniques, along with environmental conditions. In a study, Ghosh *et al.*, successfully fabricated a smart textile using merino wool/nylon polymer nanocomposites, showcasing its potential to be an advanced microwave absorber and soft touch sensor. The dip-covered textiles exhibit excellent electrical conductivity till 90.5 Scm^{-1} in conjugation with adjustable electromagnetic interference screening efficacy reaching 73.8 dB in the X band. Moreover, the in-plane thermal conductivity of 0.81 W/mK was reported to have a minimal dimension of 0.84 mm. The innovative fabric holds promise as an adjustable single item for use in smart clothing and handy advanced electronics, addressing the increasing concern of electromagnetic pollution [38].

Self-healing Ability

One of the defining features of smart polymer nanocomposites is their remarkable self-healing capability, which allows them to recover automatically when damaged. Based on the self-healing mechanism employed, these materials are categorized into extrinsic and intrinsic. Extrinsic self-healing mechanisms depend on an external repairing agent, typically as capsules or delivered through vascular structures. In the design of extrinsic self-healing materials, careful consideration is being given to embedding the repairing agent and catalyst within the matrix. These components work together to undergo the repairing process, increasing the material's resilience and extending its functional lifespan. The two commonly employed approaches for preparing materials with extrinsic self-healing capabilities are microencapsulation and microvascular structure designs. In either methodology, the self-healing procedure is triggered by impairment externally or internally within the vascular networks or the capsules. The capsules typically consist of materials such as poly (urea- formaldehyde), polyurethane, poly (melamine-urea-formaldehyde), and poly (melamine-formaldehyde), among others. A nanocomposite comprising single-walled carbon nanotubes (SWCNTs) was prepared along with 5-ethylidene-2-norbornene (5E2N) that was reacted with ruthenium Grubbs catalyst (RGC). This composite was infused within the hollow passage of the epoxy resin, thereby serving as a repairing agent in a microvascular self-repairing strategy. In order to, evaluate the self-repairing capability a cavity was formed by releasing a mass into the sample. Then, the ruptured specimen underwent thermal repairing by being heated to 60°C for 15 minutes, during which the cavity was loaded with the healing agent *i.e.*, 5E2N/SWCNT nanocomposite. The epoxy resin was observed to regain its structural integrity within 30 minutes. This creative method holds promise for enhancing the durability and longevity of materials in different applications, particularly where rapid self-repair mechanisms are required [39]. Similarly, in another study, a capsule-based approach was employed, incorporating 2.5% carbon nanotubes (CNT) added with ethyl phenylacetate (EPA) as a repairing agent. This technique utilizes CNTs along with the healing agents to enhance both mechanical and electrical healing efficiency. By adding CNTs into the healing agent, the composite material gains improved mechanical strength and durability, allowing it to effectively repair damage and restore its structure. Moreover, the presence of CNTs enhances electrical healing efficiency, providing an added benefit for applications where electrical conductivity is crucial. This capsule-based approach offers an innovative pathway in the development of new-generation self-healing materials holding outstanding mechanical and electrical properties [40]. Intrinsic self-healing materials possess the unique ability to undergo multiple self-healing processes through dynamic non-covalent relations (*i.e.*, H-bonds and π–π layering) as well as dynamic covalent bonds (including urea bonds and Diels–Alder bonds). Self-healing systems derived from dynamic covalent bonds offer distinct benefits including higher durability, enhanced thermal firmness, and increased solvent resistance in contrast to non-covalent interactions [41].

Biocompatibility

The biocompatible nature of smart polymer nanocomposites is fundamental for their application in various biomedical fields, particularly in drug delivery. The primary requirement for biomaterials intended for use in drug delivery systems is their biocompatibility, which ensures that they can be metabolized without causing any harmful effects. Such systems must possess the ability to cross the body's anatomical barriers, regulate the concentration of the drug over time, and deliver the active molecule precisely to the target site of action. Recently, there has been escalating curiosity in the advancement of smart composites or nanocomposites with enhanced biocompatibility, reflecting the demand for safe and effective drug delivery solutions. For example, a recent study reported the preparation of intelligent and bio-friendly brush copolymer attached to hybrid magnetic graphene oxide nanostructure with respect to drug release. The primary objective of the study was to enhance the medicinal efficiency of doxorubicin (DOX), a widely used anticancer agent, by encapsulating it within the nanostructure. This novel approach holds promise for enhancing targeted drug delivery and minimizing adverse effects in relation to conventional chemotherapy treatments [42]. In a different study, bio-friendly nanocomposite hydrogels were developed utilizing chitosan (CH), poly(vinyl alcohol) (PVA), oleo polyol, and fumed silica (SiO_2) by free radical polymerization, with the aim of facilitating anti-cancer drug delivery. The biocompatibility and cytotoxicity tests were done through MTT assays conducted on human embryonic kidney (HEK-293) and human colon (HCT116) cancer cell lines over a period of 48 hours. Cisplatin-filled nanocomposite hydrogels demonstrated controlled release in contrast with free cisplatin at both acidic and basic conditions. *In vitro* cytotoxicity tests of the cisplatin-filled hydrogels utilizing the HCT116 cancer cell line revealed their ability to effectively restrain the development of HCT116 cancer cells. The results suggest that the above nanocomposites have potential future applications in the development of topical drug delivery systems [43].

APPLICATION IN DRUG DELIVERY SYSTEM

Drug delivery systems play an important role in assuring reliable and effective delivery of pharmacologically active species to targeted areas within the body, thereby minimizing potential side effects and increasing treatment outcomes. A smart polymer nanocomposite has evolved as a favorable candidate for drug delivery applications owing to its distinct features and capabilities. These systems have the ability to react to various stimuli like pH, temperature, light, or specific biomolecules, allowing for precise control over drug release kinetics and targeting. These smart materials offer the capability to encapsulate and deliver drugs with improved efficiency and precision. There are various applications of smart polymer nanocomposites in drug delivery applications including targeted drug delivery, controlled release, stimuli-responsive release, combination therapy, localized therapy, responsive imaging and therapy, and implantable devices to name a few. Fig. (**1**) shows various applications related to these smart materials in drug delivery systems. Among the different types of polymeric nanocomposites used in drug delivery systems, layered double hydroxides (LDH) and polylactic acid have garnered significant attention. These materials have been successfully loaded with various drugs, including the anti-inflammatory agent ibuprofen, for targeted delivery to specific regions within the body. By harnessing the unique properties of LDH and polylactic acid, (*i.e.*, biocompatibility, tunable drug release kinetics, and reacting to external stimuli), these nanocomposites hold immense potential in the advancement of innovative drug delivery having excellent medicinal efficiency and minimum side effects. This section will explore a range of studies conducted in the field of smart materials, specifically focusing on their applications in drug delivery.

A study was conducted to synthesize dual sensitive and smart nanocomposite comprising poly(ethylene glycol) (PEG), poly (methacrylic acid) (PMA), poly (N-isopropyl acrylamide) (PNIPAM) and Fe_3O_4 nanoparticles utilizing reversible addition-fragmentation chain transfer (RAFT) polymerization. The drug release was dependent on these double-sensitive units, utilizing a pH- and thermo-sensitive triblock terpolymer as a drug vehicle. The terpolymer was prepared *via* RAFT polymerization of N-isopropyl acrylamide and methacrylic acid monomers with a PEG-RAFT agent. The Fe_3O_4 magnetic nanoparticles were prepared by co-precipitation of Fe salts. The dual sensitive behavior and self-organization of the triblock terpolymers were studied in an aqueous medium by UV–vis transmittance and dynamic light scattering. The average size of the micelles was found to be 170, 125, and 30 nm. This terpolymer was applied as a chemotherapy drug vehicle, with doxorubicin serving as a drug model. Drug discharge studies were conducted at 37 °C and 42 °C against 5.8 and 7.4 pH levels. *In vitro*, cytotoxicity analysis of free doxorubicin and drug-filled magnetic nanoparticles demonstrated biocompatibility and lack of toxicity. Additionally, the IC50 values of doxorubicin-filled terpolymer, as determined by the 3-(4,5-dimethylthiazol-2-Yl)-2,5-diphenyltetrazolium bromide (MTT) test, suggested the potential suitability of these systems for cancer treatment [44].

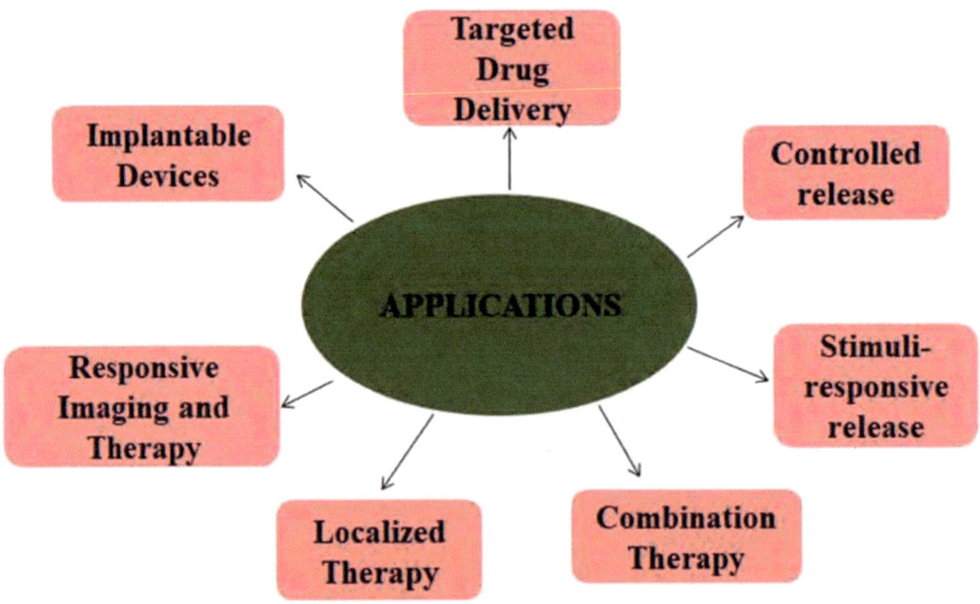

Fig. (1). Smart polymer nanocomposites in drug delivery applications.

In another study, a new versatile nanocarrier was fabricated, in conjugation with two potent anticancer agents, Doxorubicin (DOX) and methotrexate (MTX). It was demonstrated that the conjugation of DOX and MTX with the nanocarrier resulted in a controlled discharge of each of the drugs in a pH-activated approach. The co-administration of DOX with MTX showed synergistic anticancer effects on MCF7 cell lines confirmed by quantitative real-time polymerase chain reaction (qRT-PCR) and MTT assay tests. The evaluation of antitumor activity was done using human breast adenocarcinoma MCF7 cell lines as the target cells, while cytotoxicity of the nanocarrier was assessed by the MTT assay. The dual anti-cancer drug release system exhibited great antibacterial effects, and outstanding stimuli-reacting co-drug discharge capabilities with exceptional anti-tumoral efficacy. Hence, the above innovative drug carrier enables antibacterial and anti-cancer therapeutic methods to progress smoothly, thereby potentially enhancing the field of nanomedicine [45].

The utilization of nanocarriers for drug delivery to tumors has gained considerable attention, due to their exceptional drug-loading capacity. With a focus on the distinct properties of the tumor microenvironment, like its acidic pH, there has been increasing interest in the fabrication of smart nanocarriers that are capable of responding dynamically to these conditions. Such kinds of responsive nanocarriers have the potential to exert a profound influence on the speed and degree of drug discharge within the tumor microenvironment, thereby increasing the efficiency of tumor treatment strategies. In a recent study, researchers have focused on the combination of multiwall carbon nanotubes (MWCNTs) and pH-responsive nanogels to develop a versatile magneto/pH-sensitive nano-hybrid system for the delivery of DOX. The physical and chemical properties of the nanocarrier, along with the drug-filling efficacy and drug discharge behavior were evaluated. As per reports, modified CNTs exhibited lower pH response in acidic environments. However, chitosan-filled magnetic nanocomposites showed better pH responsiveness, with increased drug discharge rates (until one week) in contrast to nanocomposite systems lacking chitosan. Furthermore, Live/Dead assay experiments were conducted on U-87 glioblastoma cell lines open to DOX release supernatant at various time durations. The result clearly showed a significant inhibitory impact of DOX supernatant on cancer cell proliferation, highlighting the promising therapeutic efficacy of the synthesized nano-hybrid system in cancer treatment [46].

Lately, various approaches have been explored to develop stimuli-reacting quantum dots (QDs)-derived fluorescent probe nano gel composites for use as drug carriers, offering concurrent imaging and biosensing capabilities. One such method involves the incorporation of surface-functionalized QDs amidst the polymer structure of nanogels. Hasegawa *et al.* developed a technique wherein they generated monodisperse hybrid nanoparticles measuring 38 nm in diameter. This was obtained by blending protein-covered QDs and amino-functionalized cholesterol-loaded pullulan nano gels (CHPNH2-QD) into phosphate-buffered saline (PBS) solution for a duration of 30 minutes at room temperature. Even though the addition of QDs into unaltered, neutral cholesterol-loaded pullulan (CHP) nano gels was inhibited, the electrostatic interaction between negatively charged QDs and positively charged amino-comprising nano gels facilitated the construction of the hybrid nano gel system. Additionally, the QDs@CHPNH2 nanogel complex exhibited no agglomeration following its internalization into several human cells. This approach might be promising in the formation of advanced drug carrier systems with increased imaging as well as biosensing capacities [47].

A group of researchers synthesized and investigated a nano-scale graphene polymer composite for an oral carrier of DOX. This system comprised a DOX-filled nanocomposite containing grapheneoxide/poly(2-hydroxyethy lmethacrylate)-g-poly(lactide)-b-polyethyleneglycol-b-poly (2hydroxyethy lmethacrylate)-g-poly(lactide), formed by reversible addition fragmentation chain (RAFT) and ring open polymerization (ROP). It was reported that DOX was effectively incorporated within the nanocomposite structure with a smaller particle size of 51 nm and encapsulation efficiency of 82% ±1.12%. The study indicated that DOX has been connected to the graphene surface by hydrophobic and π-π interactions. After 72 hours, drug discharge occurred at both neutral and acidic pH, with percentages up to 24.7% and 41.2% respectively. The cytotoxicity tests conducted on 4T1 murine breast cancer cells confirmed the anticancer action of the DOX@GO nanocomposite. Overall, the nanocomposite showed potential results for breast cancer applications [48].

A significant area of focus and notable advancements has stemmed from efforts to discover novel pharmaceuticals based on the exoskeleton of marine crustacean residues, aimed at lowering environmental contaminants. In the study, higher molecular weight chitosan has emerged as a key component in the preparation of intelligent 3D cross-linked structure polymers. Here, super absorbent chitosan–ZnO nanocomposite hydrogels have been prepared using a terminated diisocyanate component, a chitosan sample (NH_2/NCO) with varying concentrations of ZnO nanoparticles ranging from 1% to 7%. The identification of this smart drug delivery system was validated by various spectroscopic methods. The distinctive physical features exhibited by the nanocomposite hydrogels in response to environmental stimuli and their porous structures have garnered significant attention. These properties are important, particularly for the development of fluorescence isothiocyanate-marked insulin-filled hydrogel nanocomposites for *in-vivo* drug delivery into the rat nasal cavity. Using confocal laser scanning microscopy, the absorption increments of the hydrogels within the rat nasal cavity were demonstrated. Moreover, the administration of these hydrogel formulations led to a notable reduction in the blood glucose amount (approximately 50-65%) for a duration of 4-5.5 hours after administration, without any cytotoxic effects [49].

CONCLUSION AND FUTURE PROSPECTS

Smart polymer nanocomposites have shown immense potential in drug delivery applications, showcasing remarkable control over drug release kinetics, targeting specific sites, and responding to external stimuli for triggered release mechanisms. These smart materials play a pivotal role in improving drug delivery efficiency by encapsulating therapeutic agents, protecting them from degradation, and facilitating controlled release profiles. They can be modified to discharge drugs as a reaction to specific stimuli like pH, temperature, enzymes, or magnetic fields, thereby minimizing off-target effects and maximizing therapeutic efficacy. Moreover, the biocompatible nature of these smart materials reduces toxicity and immunogenicity, enhancing the safety profile of drug delivery systems, and minimizing adverse effects. Furthermore, their versatility enables the encapsulation and delivery of a wide variety of therapeutic agents, from small molecules to biomacromolecules, thereby addressing a broad spectrum of medical conditions. Looking ahead, continued research efforts focusing on optimizing fabrication techniques, enhancing biodegradability, and improving targeting efficiency are anticipated to further propel the field of smart polymer nanocomposites in drug delivery, paving the way for innovative therapeutic solutions with enhanced efficacy and reduced adverse effects.

REFERENCES

[1] Namazi, H. Polymers in our daily life. *Bioimpacts,* **2017**, *7*(2), 73-74. [http://dx.doi.org/10.15171/bi.2017.09] [PMID: 28752070]

[2] Satturwar, P.M.; Fulzele, S.V.; Dorle, A.K. Biodegradation and *in vivo* biocompatibility of rosin: a natural film-forming polymer. *AAPS PharmSciTech,* **2003**, *4*(4), 434-439. [http://dx.doi.org/10.1208/pt040455] [PMID: 15198550]

[3] Belgacem, M.N.; Gandini, A., Eds. *Monomers, polymers and composites from renewable resources*; Elsevier, **2011**.

[4] Brady, J.; Dürig, T.; Lee, P.I.; Li, J.X. Polymer properties and characterization.*Developing solid oral dosage forms*; Academic Press, **2017**, pp. 181-223. [http://dx.doi.org/10.1016/B978-0-12-802447-8.00007-8]

[5] Russo, S.; Valero, A.; Valero, A.; Iglesias-Émbil, M. Exergy-based assessment of polymers production and recycling: an application to the automotive sector. *Energies,* **2021**, *14*(2), 363. [http://dx.doi.org/10.3390/en14020363]

[6] Chow, W.S.; Mohd Ishak, Z.A. Smart polymer nanocomposites: A review. *Express Polym. Lett.,* **2020**, *14*(5), 416-435. [http://dx.doi.org/10.3144/expresspolymlett.2020.35]

[7] Falahati, M.; Ahmadvand, P.; Safaee, S.; Chang, Y.C.; Lyu, Z.; Chen, R.; Li, L.; Lin, Y. Smart polymers and nanocomposites for 3D and 4D printing. *Mater. Today,* **2020**, *40*, 215-245. [http://dx.doi.org/10.1016/j.mattod.2020.06.001]

[8] Hoffman, A.S.; Stayton, P.S. Applications of "smart polymers" as biomaterials. In: *In Biomaterials Science*; Academic Press, **2020**; pp. 191-203.

[9] Aguilar, M.R.; San Román, J. Introduction to smart polymers and their applications.*Smart polymers and their applications*; Woodhead Publishing, **2019**, pp. 1-11. [http://dx.doi.org/10.1016/B978-0-08-102416-4.00001-6]

[10] Guo, Y.; Ruan, K.; Shi, X.; Yang, X.; Gu, J. Factors affecting thermal conductivities of the polymers and polymer composites: A review. *Compos. Sci. Technol.,* **2020**, *193*, 108134. [http://dx.doi.org/10.1016/j.compscitech.2020.108134]

[11] Hsissou, R.; Seghiri, R.; Benzekri, Z.; Hilali, M.; Rafik, M.; Elharfi, A. Polymer composite materials: A comprehensive review. *Compos. Struct.,* **2021**, *262*, 113640. [http://dx.doi.org/10.1016/j.compstruct.2021.113640]

[12] Schadler, L.S.; Brinson, L.C.; Sawyer, W.G. Polymer nanocomposites: A small part of the story. *J. Miner. Met. Mater. Soc.,* **2007**, *59*(3), 53-60. [http://dx.doi.org/10.1007/s11837-007-0040-5]

[13] Wells, C.M.; Harris, M.; Choi, L.; Murali, V.P.; Guerra, F.D.; Jennings, J.A. Stimuli-Responsive Drug Release from Smart Polymers. *J. Funct. Biomater.,* **2019**, *10*(3), 34. [http://dx.doi.org/10.3390/jfb10030034] [PMID: 31370252]

[14] Sahayaraj, A.F.; Dhamotharan, S.; Sandeep, D.; Ramachandran, P.; Jenish, I.; Divakaran, D.; Suyambulingam, I.; Sanjay, M.R.; Siengchin, S. Sustainable smart polymer composite materials: a comprehensive review.*E3S Web of Conferences*; EDP Sciences, **2023**, Vol. 428, p. 02014. [http://dx.doi.org/10.1051/e3sconf/202342802014]

[15] Ehrmann, G.; Ehrmann, A. 3D printing of shape memory polymers. *J. Appl. Polym. Sci.,* **2021**, *138*(34), 50847. [http://dx.doi.org/10.1002/app.50847]

[16] Ikram, H.; Al Rashid, A.; Koç, M. Additive manufacturing of smart polymeric composites: Literature review and future perspectives. *Polym. Compos.,* **2022**, *43*(9), 6355-6380. [http://dx.doi.org/10.1002/pc.26948]

[17] Liu, H.; Li, Q.; Zhang, S.; Yin, R.; Liu, X.; He, Y.; Dai, K.; Shan, C.; Guo, J.; Liu, C.; Shen, C.; Wang, X.; Wang, N.; Wang, Z.; Wei, R.; Guo, Z. Electrically conductive polymer composites for smart flexible strain sensors: a critical review. *J. Mater. Chem. C Mater. Opt. Electron. Devices,* **2018**, *6*(45), 12121-12141. [http://dx.doi.org/10.1039/C8TC04079F]

[18] Idumah, C.I. Recent advancements in self-healing polymers, polymer blends, and nanocomposites. *Polym. Polymer Compos.,* **2021**, *29*(4), 246-258. [http://dx.doi.org/10.1177/0967391120910882]

[19] Bhanushali, H.; Amrutkar, S.; Mestry, S.; Mhaske, S.T. Shape memory polymer nanocomposite: a review on structure–property relationship. *Polym. Bull.,* **2022**, *79*(6), 3437-3493. [http://dx.doi.org/10.1007/s00289-021-03686-x]

[20] Zhao, L.; Jiang, B.; Huang, Y. Functionalized graphene-reinforced polysiloxanenanocomposite with improved mechanical performance and efficient healing properties. *Journal of Applied Polymer Science,* **2019**, *136*, 47725/1-47725/11.

[21] Vera, M., Mella, C. and Urbano, B.F., Smart polymer nanocomposites: Recent advances and perspectives. *Journal of the Chilean Chemical Society,* **2020**, *65*(4), 4973-4981.

[22] Barkhordari, S.; Yadollahi, M.; Namazi, H. pH sensitive nanocomposite hydrogel beads based on carboxymethyl cellulose/layered double hydroxide as drug delivery systems. *J. Polym. Res.,* **2014**, *21*(6), 454. [http://dx.doi.org/10.1007/s10965-014-0454-z]

[23] Bellingeri, R.; Mulko, L.; Molina, M.; Picco, N.; Alustiza, F.; Grosso, C.; Vivas, A.; Acevedo, D.F.; Barbero, C.A. Nanocomposites based on pH-sensitive hydrogels and chitosan decorated carbon nanotubes with antibacterial properties. *Mater. Sci. Eng. C,* **2018**, *90*, 461-467. [http://dx.doi.org/10.1016/j.msec.2018.04.090] [PMID: 29853114]

[24] Liang, L.; Liu, J.; Gong, X. Thermosensitive Poly(*N* -isopropylacrylamide)−Clay Nanocomposites with Enhanced Temperature Response. *Langmuir,* **2000**, *16*(25), 9895-9899. [http://dx.doi.org/10.1021/la000270v]

[25] Nastyshyn, S.; Raczkowska, J.; Stetsyshyn, Y.; Orzechowska, B.; Bernasik, A.; Shymborska, Y.; Brzychczy-Włoch, M.; Gosiewski, T.; Lishchynskyi, O.; Ohar, H.; Ochońska, D.; Awsiuk, K.; Budkowski, A. Non-cytotoxic, temperature-responsive and antibacterial POEGMA based nanocomposite coatings with silver nanoparticles. *RSC Advances,* **2020**, *10*(17), 10155-10166. [http://dx.doi.org/10.1039/C9RA10874B] [PMID: 35498562]

[26] Marturano, V.; Cerruti, P.; Giamberini, M.; Tylkowski, B.; Ambrogi, V. Light-Responsive Polymer Micro- and Nano-Capsules. *Polymers (Basel),* **2016**, *9*(1), 8. [http://dx.doi.org/10.3390/polym9010008] [PMID: 30970685]

[27] Chen, Y.; Zhao, X.; Luo, C.; Shao, Y.; Yang, M.B.; Yin, B. A facile fabrication of shape memory polymer nanocomposites with fast light-response and self-healing performance. *Compos., Part A Appl. Sci. Manuf.,* **2020**, *135*, 105931. [http://dx.doi.org/10.1016/j.compositesa.2020.105931]

[28] Li, B.; Harlepp, S.; Gensbittel, V.; Wells, C.J.R.; Bringel, O.; Goetz, J.G.; Begin-Colin, S.; Tasso, M.; Begin, D.; Mertz, D. Near infra-red light responsive carbon nanotubes@mesoporous silica for photothermia and drug delivery to cancer cells. *Mater. Today Chem.,* **2020**, *17*, 100308. [http://dx.doi.org/10.1016/j.mtchem.2020.100308]

[29] Komarov, P.V.; Khalatur, P.G.; Khokhlov, A.R. Magnetoresponsive smart nanocomposites with highly cross-linked polymer matrix.Polymer Advances Technology, 1

[30] 30Zhang, K., Wang, M., Shi, Y-D., Chen, Y-F., Zeng, J-B., & Huang, J. Magnetic responsive polymer nanocomposites with *in–situ* tunable anisotropy by magnetic self-organization. *Chemistry Select,* **2016**, *1*, 5542-5546.

[31] Wang, L.; Razzaq, M.Y.; Rudolph, T.; Heuchel, M.; Nöchel, U.; Mansfeld, U.; Jiang, Y.; Gould, O.E.C.; Behl, M.; Kratz, K.; Lendlein, A. Reprogrammable, magnetically controlled polymeric nanocomposite actuators. *Mater. Horiz.,* **2018**, *5*(5), 861-867. [http://dx.doi.org/10.1039/C8MH00266E]

[32] Xu, B.; Zhang, L.; Pei, Y.T.; Luo, J.K.; Tao, S.W.; De Hosson, J.T.M.; Fu, Y.Q. Electro-Responsive Polystyrene Shape Memory Polymer Nanocomposites. *Nanosci. Nanotechnol. Lett.,* **2012**, *4*(8), 814-820. [http://dx.doi.org/10.1166/nnl.2012.1393]

[33] Chandran, P.R.; Sandhyarani, N. An electric field responsive drug delivery system based on chitosan–gold nanocomposites for site specific and controlled delivery of 5-fluorouracil. *RSC Advances,* **2014**, *4*(85), 44922-44929. [http://dx.doi.org/10.1039/C4RA07551J]

[34] Yang, C.; Liu, Z.; Chen, C.; Shi, K.; Zhang, L.; Ju, X.J.; Wang, W.; Xie, R.; Chu, L.Y. Reduced graphene oxide-containing smart hydrogels with excellent electro-response and mechanical properties for soft actuators. *ACS Appl. Mater. Interfaces,* **2017**, *9*(18), 15758-15767. [http://dx.doi.org/10.1021/acsami.7b01710] [PMID: 28425695]

[35] Vera, M.; Mella, C.; Urbano, B.F. Smart polymer nanocomposites: Recent advances and perspectives. *J. Chil. Chem. Soc.,* **2020**, *65*(4), 4973-4981. [http://dx.doi.org/10.4067/S0717-97072020000404973]

[36] Wu, M.; Chen, J.; Huang, W.; Yan, B.; Peng, Q.; Liu, J.; Chen, L.; Zeng, H. Injectable and self-

healing nanocomposite hydrogels with ultrasensitive pH-responsiveness and tunable mechanical properties: Implications for controlled drug delivery. *Biomacromolecules,* **2020**, *21*(6), 2409-2420. [http://dx.doi.org/10.1021/acs.biomac.0c00347] [PMID: 32310635]

[37] Wang, F.; Guo, Z. Facile fabrication of ultraviolet light cured fluorinated polymer layer for smart superhydrophobic surface with excellent durability and flame retardancy. *J. Colloid Interface Sci.,* **2019**, *547*, 153-161. [http://dx.doi.org/10.1016/j.jcis.2019.03.095] [PMID: 30952077]

[38] Ghosh, S.; Nitin, B.; Remanan, S.; Bhattacharjee, Y.; Ghorai, A.; Dey, T.; Das, T.K.; Das, N.C. A multifunctional smart textile derived from merino wool/nylon polymer nanocomposites as next generation microwave absorber and soft touch sensor. *ACS Appl. Mater. Interfaces,* **2020**, *12*(15), 17988-18001. [http://dx.doi.org/10.1021/acsami.0c02566] [PMID: 32207964]

[39] Aïssa, B.; Tagziria, K.; Haddad, E.; Jamroz, W.; Loiseau, J.; Higgins, A.; Asgar-Khan, M.; Hoa, S.V.; Merle, P.G.; Therriault, D.; Rosei, F. The self-healing capability of carbon fibre composite structures subjected to hypervelocity impacts simulating orbital space debris. *ISRN Nanomater.,* **2012**, *2012*, 1-16, 16. [http://dx.doi.org/10.5402/2012/351205]

[40] Bailey, B.M.; Leterrier, Y.; Garcia, S.J.; van der Zwaag, S.; Michaud, V. Electrically conductive self-healing polymer composite coatings. *Prog. Org. Coat.,* **2015**, *85*, 189-198. [http://dx.doi.org/10.1016/j.porgcoat.2015.04.013]

[41] Thakur, V.K.; Kessler, M.R. Self-healing polymer nanocomposite materials: A review. *Polymer (Guildf.),* **2015**, *69*, 369-383. [http://dx.doi.org/10.1016/j.polymer.2015.04.086]

[42] Pooresmaeil, M.; Namazi, H. Fabrication of a smart and biocompatible brush copolymer decorated on magnetic graphene oxide hybrid nanostructure for drug delivery application. *Eur. Polym. J.,* **2021**, *142*, 110126. [http://dx.doi.org/10.1016/j.eurpolymj.2020.110126]

[43] Kouser, R.; Vashist, A.; Zafaryab, M.; Rizvi, M.A.; Ahmad, S. pH-responsive biocompatible nanocomposite hydrogels for therapeutic drug delivery. *ACS Appl. Bio Mater.,* **2018**, *1*(6), 1810-1822. [http://dx.doi.org/10.1021/acsabm.8b00260] [PMID: 34996282]

[44] Ahmadkhani, L.; Akbarzadeh, A.; Abbasian, M. Development and characterization dual responsive magnetic nanocomposites for targeted drug delivery systems. *Artif. Cells Nanomed. Biotechnol.,* **2018**, *46*(5), 1052-1063. [http://dx.doi.org/10.1080/21691401.2017.1360323] [PMID: 28754064]

[45] Zakerzadeh, E.; Alizadeh, E.; Samadi Kafil, H.; Mohammad Hassanzadeh, A.; Salehi, R.; Mahkam, M. Novel antibacterial polymeric nanocomposite for smart co-delivery of anticancer drugs. *Artif. Cells Nanomed. Biotechnol.,* **2017**, *45*(8), 1509-1520. [http://dx.doi.org/10.1080/21691401.2016.1260576] [PMID: 27899033]

[46] Seyfoori, A.; Sarfarazijami, S.; Seyyed Ebrahimi, S.A. pH-responsive carbon nanotube-based hybrid nanogels as the smart anticancer drug carrier. *Artif. Cells Nanomed. Biotechnol.,* **2019**, *47*(1), 1437-1443. [http://dx.doi.org/10.1080/21691401.2019.1596939] [PMID: 30991848]

[47] Hasegawa, U.; Nomura, S.M.; Kaul, S.C.; Hirano, T.; Akiyoshi, K. Nanogel-quantum dot hybrid nanoparticles for live cell imaging. *Biochem. Biophys. Res. Commun.,* **2005**, *331*(4), 917-921. [http://dx.doi.org/10.1016/j.bbrc.2005.03.228] [PMID: 15882965]

[48] Ghamkhari, A.; Abbaspour-Ravasjani, S.; Talebi, M.; Hamishehkar, H.; Hamblin, M.R. Development of a graphene oxide-poly lactide nanocomposite as a Smart Drug Delivery System. *Int. J. Biol. Macromol.,* **2021**, *169*, 521-531. [http://dx.doi.org/10.1016/j.ijbiomac.2020.12.084] [PMID: 33340628]

[49] El-Mekawy, R.E.; Jassas, R.S. Recent trends in smart and flexible three-dimensional cross-linked polymers: synthesis of chitosan–ZnO nanocomposite hydrogels for insulin drug delivery. *MedChemComm,* **2017**, *8*(5), 897-906.
[http://dx.doi.org/10.1039/C6MD00580B] [PMID: 30108805]

CHAPTER 6

Smart Polymeric Nanocomposites for Water Purification

Tumpa Paul[1,*] and **Rajkumar Gogoi**[2]

[1] *Department of Chemistry, Darrang College, Tezpur-784001, India*

[2] *Department of Physics and Astronomy, University of Manchester, Manchester, United Kingdom*

Abstract: The process of water treatment is aimed at reducing or removing contaminants either chemical or biological in water for its utilization in safe drinking or any other essential purposes. Polymer nanocomposites that show responsiveness to external stimulants play a vital role in this process and purify water by actuating stimuli and by reviving the polymers successively. In response to environmental changes, these polymer nanocomposites can firmly undergo physical as well as chemical characteristic changes like shrinkage, and expansion, or show alterations in optical and electrical features on the basis of the induced stimulants. So far, numerous polymeric compounds that are sensitive to light, heat, CO_2, *etc.* including prefabricated nano polymers, nanoparticles, *etc.* have been employed in water treatment. This book chapter emphasizes the efficacy of different smart polymeric nanocomposites in removing harmful pollutants including toxic ions, water-soluble ions, and harmful organic colorants in the water treatment process.

Keywords: Dye, Heavy metal ions, Stimuli-responsive polymer, Water purification.

INTRODUCTION

In recent times, more emphasis has been given to research, to develop smart polymers that are sensitive to both internal and external stimuli [1 - 7]. These responsive polymers are counterfeit and produced with customised qualities so that they can exhibit preferred responsiveness to specified impulses [8]. In this book chapter, we have focused on the application of various responsive materials in water treatment, both domestic and wastewater. With the emergence of human progress, remarkable improvements in industry, and advancement in the agriculture sector, unplanned human activities are slowly causing water contamination, resulting in scarcity of fresh water. The toxic chemicals including

* **Corresponding author Tumpa Paul:** Department of Chemistry, Darrang College, Tezpur-784001, India; E-mail: tumpapaul2406@gmail.com

heavy metal ions, organic dyes, and other pollutants from industry significantly damage water bodies. The need for a permanent solution to overcome this serious issue has led to the development of many outstanding technologies that involve liquid-liquid partitioning, ultra or microfiltration, use of zeolites, resins, and other adsorbents, use of chemicals for easy sedimentation, electrochemical technique, *etc* [9 - 11]. Among them, adsorption and membrane filtration attract researchers because of their highly effectual and specific nature in pollutant exclusion, the need for minimum energy, maximum outturn, easy handling, and trouble-free expansion [11, 12]. Past years have endorsed the tremendous application of smart substances mainly nanopolymers in wastewater refinement.

These polymeric actuators are seen to be highly effective in the purification of wastewater owing to their magnificent characteristics under divergent circumstances such as heat, radiation, pH, CO_2, magnetic field, and electric field. The introduction of these materials gives a new dimension to the water treatment process, as the purification process can be governed just by changing the nature of the applied stimuli. Also, these smart polymers are self-cleanable and do not have any fouling characteristics, which leads to an increase in their functional ability and are thus cost-effective. Herein, we discuss the advancement of such delicate materials, their fabrication, features, as well as application in aquatic purification until now.

Classifications of Smart Actuators Used in Water Treatment

Stimuli-responsive materials employed in aquatic remediation are either alert to particular impulses like heat, pH, UV-visible light, energy fields, electrolytes, and ionic compounds, *etc*., or numerous impetuses, working simultaneously, otherwise in turn. This section will discuss various responsive materials that have been applied till now for the cleansing of water.

pH-sensitive Polymers

Various scientific methods, such as physical adsorption [13], oxidation [14], bioremediation [15], and stimuli-responsive materials [16], *etc*., have been employed over the past decades to remove toxic materials like heavy metals, inorganic impurities, harmful dyes, *etc*., from groundwater as well as surface water. Among the various stimuli-responsive materials, the pH-responsive materials are widely studied for water treatment. With the variation of the pH, these polymers undergo modification in their exterior action, solubility, and the sequence of the components. Basically, in an acidic medium, these responsive materials act as cationic polymers, and in the basic medium, they behave as anionic polymers, indicating their different self-assembly nature, which depends on their structures. The ionic attraction between the counter ions and dyes that are absorbed by these polymers, which alters the pH of the state, is the key factor of their high removal efficiency for these harmful chemicals.

pH-Sensitive Cationic Smart Materials

Efficient removal of NO_3^- from the water was studied by Zahra Abousalman-Rezvani and co-workers [17], who synthesized grafting light, temperature, and CO_2-responsive copolymers from cellulose nanocrystals (CNC)-grafted and free block copolymers of 2-dimethylaminoethyl methacrylate (DMAEMA) and coumarin by atom transfer radical polymerization. The structural and thermal characterizations were observed by Fourier-transform infrared spectroscopy, proton nuclear magnetic resonance, and thermogravimetric analysis. With the increase in pH, the LCST of the polymer decreased. Bubbling of CO_2 through an aqueous solution results in protonation of the amine group of the polymer, thus adsorption of the ions increases, causing LCST at higher temperatures. Deprotonation of the polymer was observed on inserting N_2 gas. Yan Qin *et al.* studied [18] the removal of methylene blue (a-MB), a negatively charged acid dye, from water with high efficiency using pH-responsive ammonium-functionalized hollow polymer particles (HPP-NH_3^+) with a high density of ammonium groups in the shell. Remarkable pH-dependent equilibrium adsorption was shown by HPP-NH^{3+} polymers that increased from 59 mg/g to 449 mg/g with a decrease in the pH from 9 to 2. The reason behind the pH responsiveness of the polymers is the presence of the dual functional group (ammonium and carboxyl groups). HPP-NH^{3+} exhibited low adsorption capacity for b-MB while for a-MB the case was reversed, indicating its selective nature for acidic dyes. The adsorption equilibrium was attained at 15 mins with an adsorption capacity of 194 mg/g, having an initial a-Mb concentration of 200mg/L. The polymers can be easily recovered at basic solution pH = 10 and can be reused after 5 cycles with 98% dye removal efficiency.

Nowadays, biopolymer-based adsorbents draw more attention in the purification of aquatic systems as they are inexpensive and environmentally safe [19]. Among them, pH-responsive biopolymers are investigated by most researchers for the removal of harmful dyes from water. Bincheng Xu and co-workers reported a novel chitosan-based magnetic adsorbent [poly(2-(dimethylamino)ethyl methacrylate) grafted magnetic chitosan microspheres, GMCMs] and their efficient adsorption of dyes that are negatively charged, such as Responsive Blue 19 and Acid Green 25. The remarkable adsorption capacity of GMCMs was found to be at pH 2 and at a contact time of 90 min due to complete protonation of the grafted PDMAEMA chains, and they became positively charged electrolytes. Along with electrostatic attraction, the attractive force *i.e.*, hydrogen bond, amid the polymers and the colorants also increased its adsorption capacity. The polymer was deprotonated and shrank above pH 8 and easily regenerated. The adsorption efficiency of this polymer was found to be 90%, even with its utilization up to 5 terms. The absorptivity of cup-shaped polylactide (PLA)/poly[methyl methacrylate-co-2-(2-bromopropionyloxy) ethyl methacrylate] (poly(MMA-co-BEMA)) (75/25) composite particles (PDMAEMA brush-modified particles) for hydrophilic Rhodamine B was investigated by Ifra, Sampa Saha [20]. Their observation revealed that higher adsorption of the dye occurred by brush-modified PDMAEMA at pH 7 due to strong electrostatic attraction between negatively charged Rhodamine B, because of more carboxylate ions and positively charged brush PDMAEMA, showing the pH responsiveness and dye selectivity of the polymer. The maximum adsorption of the dye by the was found to be 79.2mg/g and the time to reach equilibrium was less than 12 hours.

A study on the successful use of pH-sensitive poly(2-(dimethylamino)ethyl methacrylate)/nanocrystalline cellulose hydrogel for anionic methyl orange (MO) elimination was accomplished by Seyedeh-Arefeh Safavi-Mirmahalleh and colleagues [21]. The properties of the nanocomposite were analyzed using Fourier-transform infrared spectroscopy (FTIR), dynamic light scattering (DLS), and field emission scanning electron microscopy (FE-SEM). At higher pH values, the adsorption capacity of hydrogels containing 2 wt. % NCCs were decreased. Yixing Duan *et al.* synthesized polyamidoamine dendrimer gel (PAMAM) with multiple amine groups as a super adsorbent and studied the adsorption capacity for tartrazine (TTZ) and methyl orange (MO), negatively charged dyes [22]. The maximum adsorption capacity of PAMAM gel towards MO and TTZ was as high as 680.2 mg g^{-1} and 689.7 mg g^{-1} respectively. With an increase in initial dye concentration and pH, the adsorption rate decreased while it was enhanced with an increase in temperature. The presence of NaCl first reduced the uptake capability of the hydrogel and then enhanced it with an increase in the concentration of the salt. Amputation capability of the hydrogel of more than 81.0% was maintained despite the reuse of up to five times and could reach 550.9mg g^{-1} and 558.7 mg g^{-1} for MO and TTZ dye, respectively. Synthesis and application of coumarin-containing multi-responsive CNC-grafted and free copolymers (PDC5010 and C-PDC5010) for the elimination of NO_3^- ions from wastewater were examined by Zahra Abousalman-Rezvani and co-workers [23]. Maximum adsorption of NO_3^- ions by positively charged sites of the polymer was found at lower pH and purging CO_2 to the process. Also, CNC-grafted copolymers showed char residues of 16.7 and 8.3%, respectively, for the 30:5 and 50:10 monomer ratios. Enhancement of ion removal was observed on increasing the PDMAEMA block length. The polymers can be regenerated by applying UV light of about 225 nm and inserting N_2 gas that breaks the network of the polymer. The CNC-grafted polymers had lower adsorption than free polymers.

Lin Yang *et al.* studied the selective adsorption of anionic dyes such as methyl orange (MO), methyl blue (MB), and naphthol green B (NGB), using novel CO_2-responsive cotton as an eco-friendly adsorbent derived from poly (4-acryloyloxybenzophenone-co-2-(dimethyl amino) ethyl methacrylate) (PCC) [24]. The polymer displayed the highest adsorption capability for the dyes and speed to reach equilibrium. It showed maximum adsorption capacities of 1785.71 mg g^{-1} for methyl orange (MO), 1108.65 mg g^{-1} for methyl blue (MB), and 1315.79 mg g^{-1} for naphthol green B (NGB), following the adsorption equilibrium times of 5 min for MO, 3 min for MB, and 4 min for NGB. On increasing the concentration of CO_2 and lowering the pH to 4.18, the adsorption of MO increased from 16.8% to 100%. Removal of anionic dyes improves with the enhancement of the temperature because of the increase in the movement of anionic dye molecules while supplying sufficient energy to enhance the interactions between the adsorption binding sites of the PCC-3 and the anionic dye molecules. The polymer could be easily regenerated simply by heating in aqueous solution.

Anionic pH-sensitive Polymers

Hasina Kasgoz and colleagues [25] explored the adsorption capability of a novel type of hydrogel-clay nanocomposite based on the acrylamide (AAm)- 2-acrylamido2-methylpropane sulfonic acid (AMPS) sodium salt and clay (AAm-AMPSNa/clay hydrogel) for heavy metal ions (Cu (II), Cd (II), Pb (II)) from water in both competitive and non-competitive conditions. The highest adsorption of about 0.97, 0.98, and 0.89 mmol g^{-1} of Cu (II), Cd (II), and Pb (II), respectively, by the composites was found to be with 10%(wt) of clay content. The transition metal speciation, solubility, and charge of the functional group of the hydrogel are affected by changes in pH. With the increase in the pH (3-4.5), the adsorption capacity of the HC-1.1 sample also increased for all three metal ions due to the dissociation of the sulfonate groups. At a pH value of 3, the adsorption of Pb (II) was more than that of Cu (II) ions. The efficient desorption of metal ions from the hydrogel composite was found to be more than 97%. The successful removal of heavy metal ions like Cd (II), Cu (II), and Fe (III) ions [26] from their aqueous solution by using anionic acrylamide-based hydrogels (AM/AMPS) was examined by Ayman M. Atta *et al.* Swelling capacity of the hydrogels increased with an increase in the pH up to 12. Maximum adsorption of Cd (II) and Cu (II) was observed at pH 5, and for Fe (III) it was at pH 2.5. An increase in pH favored the development of metal ion composites due to a decrease in the synergic relationship between H^+ and resins. Depending on the polarity, electronic configuration, ionic radius, *etc.*, and also on the nature of interaction with the functional groups of the hydrogel, the affinity order of the metal ions for the hydrogels was found to be Fe (III) $<$ Cu (II) $<$ Cd (II). With increasing concentration of the AMPS, the adsorption capacity of the hydrogel increased. The recovery rate of the hydrogel was about 90% after 3 cycles.

Many experiments have been performed till now for the elimination of lead, a heavy metal ion, from water for its toxic and carcinogenic nature. Elizabeth Ramírez and co-workers [27] used 4-vinyl pyridine-grafted poly(acrylic acid) hydrogels (net-PAAc-g-4VP) for the removal of Pb (II) from its aqueous solution. The maximum adsorption of Pb (II) ions by the hydrogel was about 90% at pH 4 due to weaker electrostatic repulsive force; the ions were easily attached to the surface of the polymer. The adsorption of Pb (II) ions increased rapidly in the first 8 h and then the rate slightly decreased. The results revealed that with 26.74% grafting of the hydrogel per g the maximum removal of Pb (II) ions occurred, which was 117.9 mg. Zhengkui Li *et al* [28] researched the removal of heavy metal ions (Cu (II), Fe (III), Cr (III), Cd (II), and Pb (II)) from wastewater using a novel HEA/AMPS copolymer hydrogel. The characterization of the prepared copolymer hydrogels was done by scanning electron microscopy, Fourier-transformed infrared spectra, thermal gravimetric analysis, and X-ray photoelectron spectroscopy. The optimal pH and temperature regarding all the ions were 6 and 15°C, respectively. The adsorption rate of the hydrogel for the various ions was found to be $Cr^{3+} > Fe^{3+} > Cu^{2+} > Cd^{2+} > Pb^{2+}$. The maximum load capacity for Pb^{2+}, Cd^{2+}, Cr3+, and Fe^{3+} was 1,000 mg L^{-1}, whereas that for Cu^{2+} was 500 mg L^{-1}. The adsorption equilibrium of HEA/AMPS copolymer hydrogel was well described by a pseudo-second-order equation and the Langmuir adsorption isotherm.

The elimination of organic dyes like bisphenol A (BPA) and methylene blue (MB) by self-assembled cyclodextrin polymer (P-CD)/ferrocene-modified poly acrylic acid (PAA-Fc) was studied by Nan Hou and colleagues [29]. They observed that the highest adsorption capacity of the hydrogel material attained 23.61 mg/g for BPA and 24.64 mg/g for MB and had high pH dependence. Poor absorption adequacy for BPA in an alkaline medium of pH > 9 was manifested by the hydrogel, and for MB it was the acidic solution of pH < 5. The hydrogels removed BPA through host-guest interaction as the dye was hydrophobic and eliminated positively charged MB *via* electrostatic interaction and hydrogen bonding. The highest adsorption by the hydrogel for BPA and MB was found to be 23.61 mg/g for BPA and 19.74 mg/g for MB respectively proving itself a good adsorbent in wastewater treatment.

Large amounts of dyes released from the industries in water cause serious problems to the environment due to their strong color and low ratio of BODs to CODs. Many new experiments have been done to remove these harmful dyes from water. One of them was the adsorption of cationic methylene blue (MB) using pH-responsive poly (acrylic acid-co-vinyl sulfonic acid) aquatic gel (PAAVSA), studied by Rinki Singh and co-workers [30]. The strong affinity for the cationic dye molecule by COO^- and SO_3^- groups present in the hydrogel is the reason for its removal capacity. pH-responsive hydrogel showed various swelling rates at different pH values. The adsorption capacity of PAAVSA-I was maximum in 60 mg/l of MB dye solution and at pH higher than 7 (pH = 7, pH = 9). The hydrogel was easily regenerated at pH 1.4 and the process was repeated till the 7th cycle. These properties make the polymer pH sensitive, have substantial dye adsorption capacity in a shorter period, and are thermally stable, recyclable/reusable or cost-effective, and biocompatible/eco-friendly. The elimination of methylene blue (cationic) and indigo carmine (anionic) from aqueous solution using Karaya gum-graft-poly(dimethyl aminoethyl methacrylate) (KG-g-PDMAEMA) gel was investigated by Preetha B. K. and Vishalakshi B [31]. The results showed that the highest swelling took place at a neutral medium, *i.e.*, at pH 7, and followed the order of pH: 7.4 > 9.0 > 6.8 > 12.0 > 1.2. With the increase in temperature from 25°C to 40°C, a slight increase in swelling was observed, and further temperature increases resulted in a drastic decrease in swelling. With the increased temperature and initial concentration of the dye, the rate of adsorption increased per gram. The removal capacity of the hydrogel for MB and IC was found to be 89.28 mg/g and 101.42 mg/g, respectively, which was higher than any other polymers.

Shaikh A. Ali and colleagues [32] reported the synthesis of chelating ion-exchange resin [cross-linked di-anionic polyelectrolyte (CDAPE)] for the elimination of Pb(II) and Cu(II) ions from wastewater. The removal process followed pseudo-second-order kinetics and was well-fitted with Langmuir, Freundlich, and Temkin as well as Dubinin−Radushkevich isotherm models. The maximum adsorption of Pb^{2+} and Cu^{2+} were observed to be 3.83 and 10.1 mmol g−1 at optimum pH =4 and pH=5.3 respectively. The removal rate shown by the resin was enhanced with the raising in the preliminary amount of the ions. The resin could be reused up to 3 cycles.

Julio Sánchez and colleagues also reported the synthesis and adsorption capability of Poly(N, N-dimethyl amino ethyl methacrylate) (PDMAEMA) for eliminating Cr^{+6} ions by applying polymer-based ultrafiltration mechanism [33]. The characterization of the polymer was done with size exclusion chromatography (SEC) and proton nuclear magnetic resonance (1H NMR) spectroscopy. The maximum adsorption of Cr^{+6} through PDMAEMA using the PEUF technique showed 100% for 25 mg L^{-1} at pH = 4. The maximum percentage of Cr (VI) was eliminated when the polymer and Cr^{+6} ratio was 40:1 and 80:1. The intake of ions by the polymer was not significantly changed in the presence of interfering radicals like $H_2PO_4^-$ and SO_4^{2-}, as the ions didn't compete with Cr^{6+} for the active sites of PDMAEMA. The easy recovery of the polymer took place at pH 10 and could be reused up to nine cycles.

The efficient elimination of Ni (II) ions and cationic malachite green (MG) using carboxylate-rich magnetic chitosan flocculants (Mag@PIA-g-CS) was studied by Bingzhi Liu *et al.* [34]. Mag@PIA-g-CS exhibited its maximum sorption performance in a wide pH range (4.0-8.0 for Ni (II) and 5.0-10.0 for MG) with optimal adsorption rates of 98.3% and 87.4%, respectively. The optimal doses of Ni (II) and MG uptake were found to be 150 and 200 $mg{\cdot}L^{-1}$ for Mag@PIA-g-CS. The polymer showed rapid intake of ions and dyes due to the strong interaction of target contaminants with amino, hydroxyl, and carboxyl functional groups and the lack of internal surfaces (such as holes and crevices). The uptake capability of flocculants remained at about 84% for Ni (II) and 72% for MG after five consecutive cycles. These flocculants can be used in water treatment for their stability, reusability, and capacity for rapid elimination. The removal of Cr (III) ions with the help of novel cross-linked pH-responsive tetrapolymer (CPZA) having an amino, carboxylate, phosphonate, and sulfonate motifs was investigated by Tawfik A. Saleh *et al.* [35]. At an optimum pH of 5.5 and temperature of 296 K, CPZA showed the highest adsorption capacity of 48.5 mg/g for Cr (III) ions. The adsorption process was well described by the pseudo-second-order kinetic model. The ease of reusability and good stability made the polymer a good option for water treatment. Ultrafiltration membrane-based polymers nowadays become very popular for removing heavy metal ions from water as they can perform in a homogeneous medium.

Shaikh A. Ali and co-workers investigated the removal of Cr (III) and organic dyes, including methyl orange, Eriochrome black T, rhodamine B, methyl red, and methylene blue, from their mixture by a pH-responsive resin containing zwitterionic amino methyl phosphonate ligands and hydrophobic pendants. The highest intake of Cr^{3+} ions by this polymer occurred 100% in 20 min at pH 5, with an initial concentration of 1.0 ppm of the ions and at a temperature of 25°C. The presence of hydrophobic aromatics capable of π-π interactions in the resin was the cause of its affinity towards dyes. It showed significant proficiency in extracting rhodamine B Eriochrome black T, methyl orange, methyl red, and methylene blue with removal percentages of ≈100, ≈100, 95, 92, and 90%, respectively, attained at 30 min. Even after 3 cycles, the resin was capable of eliminating the ion and dyes with ±4% changes. Cellulose-based adsorbents are widely employed in wastewater treatment for the removal of hazardous and toxic heavy metal ions and organic dyes for their low capital cost, low energy consumption, and ease of operation. Juntao Tanga *et al.* reported the synthesis and characterization of compressible cellulose nanofibril (CNF)-based aerogels produced *via* a bioinspired strategy for Cu (II) ion and methyl orange (MO) dye removal [36].

The adsorption process was well described by the 2nd order kinetic model and the mechanism by the intra-particle diffusion model. The maximum adsorption capacity of 103.5 mg/g and 265.9 mg/g for Cu (II) and MO, by the polymer, at pH 5 and pH 4.41, respectively. At high pH, as the amounts of positively charged groups on the aerogel surface decreased, the intake capability of this polymer reduced. The eradicable efficiency of the polymer for Cu (II) exceeded over 91% and 89% respectively, after successive 4 cycles.

The adsorption method using histamine, a well-known amidine featuring an imidazole ring that not only exhibits CO_2^- and pH-sensitive characteristics but also forms metal-ligand interactions with metal ions, is considered the most environmentally friendly, effective, efficient, and economic method of waste purification among the various physicochemical treatment processes. Ngoc Bich Tran *et al.* studied the successful removal of Cu^{2+}, and Ni^{2+} by novel polyaspartamide copolymers containing histamine pendants (PHEA-HIS) [37]. The swelling degree of the PHEA-HIS gel decreased noticeably from $SR_{\sim}$ 50 g g^{-1} to SR $_{\sim}$18 g g^{-1} when CO_2 was bubbled into the solution containing the gel specimen as the imidazole ring of the histamine acted as a proton donor. The adsorption increased from 6–42.35 mg g^{-1} for Cu (II) and from 27–78 mg g^{-1} for Ni (II) in aqueous solution on increasing the initial concentration from 50 to 500 ppm and pH from 3 to 5. The successful eradication of Cu (II) ions from aqueous solution was investigated by J.J. Chen and co-workers using poly(N-isopropylacrylamide-co-acrylic acid) hydrogels (PNIPAM-co-AA) [38]. The results revealed that the maximum adsorption capacity of the polymer was 67.25 mg/g at 303 K and pH 5. The adsorption mechanism favored chemisorption and was well described by the pseudo-second-order kinetic model. Adsorption of Cu (II) ions on PNIPAM-co-AA hydrogels was found to be more thermodynamically driven.

CO_2 Responsive Polymer

CO2-responsive materials are mostly PDMAEMA-based polymers that allow rapid and complete regeneration and metal recovery by simply purging CO_2 gas into the solution under room temperature and atmospheric pressure. The adsorption of Cu(II) ions from aqueous solution by CO_2- responsive polymeric octopus made of a silsesquioxane core and poly(2-dimethylaminoethyl methacrylate) (POSS-PDMAEMA Octopus) was observed by Yu Bai and colleagues [39]. The adsorption of Cu(II) ions by the polymer occurred through the formation of coordination bonds between the dimethyl amino groups and the metal ions with a maximum capacity of 145.1mg/g. The degree of protonation of the dimethyl amino groups is estimated to be ca. 95% at pH 4.0 when adsorption of the ions took place, and deprotonation happened at pH between 4.5 and 6.5 that differ the said polymer from other adsorbents, which required pH 1 for regeneration. It becomes a great task for the researcher to develop a chemical-free, economically friendly desorption method for wastewater treatment. Recently, Songlin Fan and colleagues [40] synthesized CO_2 -responsive chitosan aerogel from CO_2-responsive poly(acrylic acid-2-(dimethylamino)ethyl methacrylate) and chitosan (CPA) by physicochemical double crosslinking for the adsorption of Cu2+ from aqueous solution. The equilibrium adsorption capacity of CPA was about 130.6 mg g^{-1} at 360 min, which was 162% higher than chitosan aerogel (CA). With enhancement of the pH of the solution, the adsorption rate increased and became maximum at pH 5.5, again during the adsorption of the Cu^{2+} ion, the pH of the solution decreased because of the facile deprotonation of the –COOH group in CPA at a high initial pH and indicating that ion exchange occurred between Cu^{2+} in solution and H^+ on the adsorbents during the adsorption process. By bubbling CO_2 in a mild condition, the desorption of CPA could be easily done, and even after 6 cycles, the adsorption capacity of CPA for Cu^{2+} still reached 70% of the initial adsorption capacity, and the desorption rate reached 75%, which made this aerogel a suitable adsorbent for the elimination of heavy metal ions from water.

Temperature-Responsive Polymer

PNIPAM polymers that have been engaged in the refining of wastewater are mainly temperature-sensitive. Modified PNIPAM is used for the elimination of heavy metal ions and dyes from aqueous solution. Modified PNIPAM, a temperature-responsive polymer, undergoes a tremendous volume alteration at its lower critical solution temperature (LCST, ~32°C) [41, 42]. So, this type of smart material has wide application as a heat-sensitive material in the membranes that are susceptible to heat [43, 44]. Herein, we will discuss the synthesis and utilization of some thermoresponsive polymers in water purification.

Xiao-Jie Ju *et al.* studied [45] the removal capacity of a novel polymer prepared by incorporating benzo-18-crown-6-acrylamide (BCAm) as a metal ion receptor into the thermo-responsive poly(N-isopropylacrylamide) (PNIPAM) hydrogel [P(NIPAM-co-BCAm)] for Pb(II) ions from water. This hydrogel showed intense adsorption of Pb(II) ions at lower temperatures, below LCST (lower critical solution temperature), whereas desorption took place above LCST, indicating the thermosensitivity of the hydrogel. With the change in the environmental temperature, the stretch-shrink configuration of the polymer network changes. Thus it has great potential application in eliminating Pb(II) ions from water just by simply changing the temperature of the environment.

Nowadays, the microgels prepared from poly (N-isopropylacrylamide) (pNIPAm) attract researchers for the elimination of hazardous dyes from wastewater as they are hydrophilic in nature, extremely penetrable, and thermosresponsive. The development of poly (N-isopropylacrylamide) microgels and also the study of their adsorption capability for the organic azo dye molecule, 4-(2-Hydroxy- 1-naphthylazo) benzenesulfonic acid sodium salt (Orange II), from aqueous solutions at both room and elevated temperature [46] was reported by Deepika Parasuraman and Michael J. Serpe. They found that the adsorption capacity of the pNIPAM-co-10AAc enhanced from 3.60% to 29.5% with the increase in the amount of the microgel up to 521 μg microgels/mL, but there was no remarkable change in the adsorption with further increase in the amount of the microgel. The extent of Orange II removal was solution temperature dependent; more Orange II was removed from water at elevated temperatures and the microgels were held at that temperature for longer durations of time. Even by increasing the cycles between high and ambient temperature, the removal of Orange II was increased. The leaking percentage of the Orange II by the microgels was found to be 25.6% from the leaking studies without the occurrence of any discharge after 50 min. Zhuang Liu and colleagues [47] developed a gating membrane with poly(N-isopropylacrylamide-co-acryloylamidobenzo-18-crown-6) (poly(NIPAM-co-AAB18C6)) copolymer chains as functional gates, [poly(NIPAM-co- AAB18C6)-grafted gating membranes] for selective detection and removal of Pb(II) ions from the environment. Removal of Pb(II) ions took place at 25°C within a treatment time of 34 hrs due to swelling of the poly(NIPAM-co-AAB18C6) chains, making it easier for the cavities of 18-crown-6 units to capture the Pb^{2+} ions from the environment. The said polymer was easily regenerated by changing the temperature to 45°C. This thermosensitivity of the polymer makes it superior to other polymers in the area of water decontamination.

A study of the absorption and discharge of a positively charged dye, orange II by microgels of N-isopropylacrylamide copolymer with 1-vinylimidazole (PNIPAM-co-Vim) was carried out by Tomasz Śliwa and associates. P(NIPAM-co-Vim) particles at pH 4 carry a positive charge, and at pH 9 have a negative charge [48]. The interaction between P(NIPAM-co-Vim) and the dye molecules depends on the electrostatic charge, which was strongly influenced by the pH. The optimum pH for maximum adsorption of dye by the polymer was 4, and the dye bonding was found to be 47% and 87% at temperatures below and above LCST, respectively. With the shifting of the pH to 9, desorption of the dye took place.

Light Responsive

Irradiation of UV/visible light can be applied for the eradication of toxic pollutants from aqueous solution by varying the features of the substances that include color, charge of the surface, polar and non-polar properties, sol-gel, and also change in the structure of the pore [49 - 52]. Kristen H. Fries *et al.* fabricated a methacrylate-based spiropyran-containing thin film sensor [Poly (MMA90-co-SPMA10)] and studied its efficiency in the identification of divalent metal ions Sn^{2+}, Cu^{2+}, Fe^{2+}, Zn^{2+}, Co^{2+}, and Ni^{2+} from a binary mixture in the presence of UV-vis light [53]. It had the capability of identifying both metal ions in a binary mixture. On immersion of the polymer in a binary mixture of two different ions (having the same six metal ions) in varying concentrations for 1 min under UV-vis light, it was found that the film solely bound with Sn^{2+} ions at any concentration ratio with Fe^{2+}, Co^{2+}, Ni^{2+}, or Zn^{2+}. With mixtures of Sn-Ni, Zn-Ni, and Fe-Ni, the SPMA copolymer film always preferentially formed complexes with the ion that was not Ni^{2+} at any concentration ratio. The film was more successful in identifying Sn^{2+} and Cu^{2+} than the other ions because it achieved 100% sensitivity for Sn^{2+} and greater than 98% sensitivity for Cu^{2+}. Both ions also had greater than 96% specificity. The film was also successful in identifying Fe^{2+} (89% specificity and 91% sensitivity). The model was moderately successful in predicting Co^{2+}, Ni^{2+}, and Zn^{2+}, although neither the sensitivity nor the specificity was less than 80% for any ion.

In recent years, the use of biopolymers, especially silk fibroin (SF) in nanocomposites, has attracted researchers, as it possesses excellent biocompatibility, optical transparency in the visible range of the electromagnetic spectrum, and facile processability. Maria E. Genovese and co-workers [54] also developed photochromic spiropyran-doped silk fibroin poly-(ethylene oxide) nanofibers (SF/PEO/SP), which combined the attractive properties and biocompatibility of silk with the photo-controllable and reversible optical, mechanical, and chemical response of the spiropyran (SP) dopants. SP has the capability to interconvert reversibly between two different isomers, namely, a colorless polar spiropyran (SP) and a purple zwitterionic merocyanine (MC), upon light irradiation. When UV-irradiated nanofibers came in contact with the acidic environment, a 90% decay of the emission signal was observed after 30 s, accompanied by a corresponding chromic variation from purple to yellow, indicating its acid responsiveness. On immersion of UV-irradiated SF/PEO/SP nanofibrous mats in ethanolic solutions of Zn^{2+}, Hg^{2+}, Pb^{2+}, and Cu^{2+} ions, a chromic variation from purple to pink or orange (metallochromism) occurs after 60 sec. The chromic changes happened due to the formation of different complexes through the chelation of each ion by the phenolate and the methoxy oxygens of MC. Along with the identification of the metal ions, it can also measure the adsorption of the ions. Cu^{2+}, Hg^{2+}, and Pb^{2+} exhibited a similar affinity with the. nanofibers, with adsorption capacities ranging from 0.012 to 0.40 mmol g^{-1} for the above-mentioned concentrations, whereas the adsorption of Zn^{2+} was lowered by approximately 50%. The nanofibers could be reused for several cycles with good sensing performance.

Suyan He *et al.* [55] described the detection of Fe^{3+} ions by the small fluorescent molecule boron-dipyrromethene (BODIPY) co-polymerized with MAA by RAFT method. The fluorescence value and the maximum absorption wavelength didn't show a noticeable change when the pH value was in the range of acidic and neutral conditions, but as it exceeded 7.53, a slight change appeared in the fluorescence value and waveform. The fluorescence of the chemical sensor gradually reduced when the concentration of Fe^{3+} increased from 0 to 15 folds of the equivalent. The sensitivity of the BODIPY-based polymer for the identification of Fe (III) ions was observed even at very low content (0.5%).

Magnet Field Responsive Polymer

Smart materials that show a significant change in properties in response to the applied magnetic field have found wide scope in the arena of sanitization of aqueous solutions. So far, research done in this field has been discussed in this section. For instance, the selective removal of toxic metal ions, Cu(II), Cd(II), Fe(II), and Pb(II), in an aqueous medium employing poly(2-acrylamido-2-metyl-1-propansulfonic acid-co-vinylimidazole), p(AMPS-c-VI) magnetic hydrogels was studied by Ozgur Ozay *et al.* [56]. According to their report, the removal efficiency of the magnetic hydrogel attained 90% just by using 0.2–0.3 g of absorbent for a 50 ppm metal ion solution. The affinity towards different metal ions varies with the types of magnetic hydrogels; *e.g.*, p(AMPS) hydrogels had better binding ability towards Fe (II) (1.14 mmol) than Cu (II) (0.255 mmol) and more selectivity for Fe (II). The release of the metal ions from the magnetic particles containing polymers took place in the presence of H_2SO_4 and the polymer can be reused up to 5 cycles. Thus, magnetic hydrogels can be used as a tool for toxic metal ion treatment in water. Yang Si and colleagues [57] studied the fabrication and utilization of magnetic polybenzoxazine-based carbon nanofibers with Fe_3O_4 inclusions with a hierarchical porous structure (Fe_3O_4@CNFs) for dye removal, mainly methylene blue (MB) and Rhodamine B (RhB) from wastewater. The complete elimination of MB and RhB occurred within 15 and 25 min, respectively. The adsorption of the dyes by the polymer is attributed to the p–p stacking between dyes and p-conjugation regions of the CNFs. Another important fact is that the aqueous suspension after adsorption could be separated easily by an external magnet.

Potential Responsive Polymer

Current research focuses on electrically switched ion exchange (ESIX) films that have tremendous utility in the purification of aqueous solutions, as it is a combination of interchanges of ions and electrochemistry to separate the metal ions in a specific and reversible manner. Zhongde Wang *et al.* [58]. also developed a novel potential-responsive ion exchange film consisting of one layer of α-zirconium phosphate (α-ZrP) ultra-thin film along with a polyaniline (PANI)-interpolated chain, a conductor that worked as a H^+ driving constituent (α-ZrP/PANI hybrid film), for heavy metal ion (Ni^{2+}) removal. The formation of steady composites by selecting specific heavy metals is the key role of α-ZrP nanofilm, an ion receiver, whereas the main task of PANI-interpolated chain is to control the pumping of H^+ at the time of altering the film potentiality in order to adjust its oxidation or reduction state. Pumping of Ni (II) ions into the film and regeneration of the composite film happened by switching the potential from 0.8 V to -0.2 V. The presence of Cd^{2+} and Pb^{2+} reduced the uptake of Ni^{2+} ions by the hybrid films. The metal uptake in the film was also dependent on the spacing between the two layers, the energy that was submerged in ions, the complexation effect, and the valency of the ions. Thus, it seems to be a better substitute for usual practices for purifying aqueous solutions, comprising heavy metals. Zhongde Wang and associates [59] reported the separation and recovery of Ce^+ ions with the help of a smart magnetic potential-responsive ion exchange nanomaterial (MPiX), c-Fe_2O_3@nickel hexacyanoferrate (NiHCF), with a three-dimensionally ordered microporous (3DOM) structure and superparamagnetism. A rapid adsorption of Ce+ (above 96.5%) was found within 10 min of the beginning, due to the presence of the open and interconnected porous channels and the large specific surface area of 3DOM and the good affinity of NiHCF toward Cs^+. With an increase in the initial concentration of the Cs^+ ion, the capture of the ion by the polymer was enhanced.

More than 98% of the adsorbed Cs^+ ions can be removed from 3DOM MPiX by switching the potential of the magnetic electrode and can be reused after 20 cycles. The fabrication and application of ultrasound-assisted poly(acrylic acid)-bentonite-FeCo (PAA-B-FeCo) hydrogel nanopolymer for the exclusion of a cationic dye, crystal violet (CV), was studied by S. R. Shirsath *et al.* [60]. The highest removal of the dye, about 87%, was obtained at an optimum temperature of 35°C and pH 11. With an increase in initial dye concentration and hydrogel, a sharp rise in the uptake of the dye was found. The reports revealed that the magnitude of abstraction of dye obtained by the mixture of hydrogel and ultrasound (97% in merely 5 h) was greater as compared to solo hydrogel adsorption (87% in 15 h).

Dual Responsive Polymer

Considerable influence over the performance of the substances under varied circumstances will be shown by the polymers that are responsive to more than one impulse, either simultaneously or autonomously [61]. The capability of temperature- and pH-sensitive poly (N-isopropylacrylamide-co-itaconic acid) [NIPAm-co-IA] for the removal of cationic dyes like Safranine T (ST), Brilliant Green (BG), and Brilliant Cresyl Blue (BCB) from aqueous solutions was investigated by Bengi O Zkahraman and fellows [62]. Swelling of the hydrogel decreased in the temperature range of 30–50°C and increased with the increase in pH up to 9, but above pH 9 the rate decreased. The adsorption capacities of the polymer followed the order BG > ST ~ BCB. Dye removal efficiency percentages of the hydrogels were in the range of 90–100% for 100 ppm dye concentration for BG and ST.

Karthika J.S. and Vishalakshi B reported a novel stimuli-responsive gellan gum - graft -poly(DMAEMA) (GG-g-Poly (DMAEMA)) hydrogel as an adsorbent for the anionic dye methyl orange (MO) [63]. The characterization of the gel was carried out by FTIR, XRD, TGA, DSC, and SEM techniques. The rise in swelling of the gel was observed at pH around 7, whereas a sharp decrease was found at a temperature around 50°C. The maximum adsorption capacity (25.8 mg g^{-1}) was attained at pH 3. Their findings indicate GG-g-Poly(DMAEMA) gels as a better adsorbent for the removal of anionic dyes from dye-contaminated water samples.

Guohong Yao and colleagues [64] investigated the capability of dual-responsive Graphene Oxide/Poly(NIPAM-co-AA) hydrogel (GO/poly(NIPAM-co-AA) (PNA) to eliminate cationic dyes, rhodamine B (RB). The characterization of the hydrogel was examined by Fourier transform infrared spectroscopy, X-ray photoelectron spectroscopy, Raman spectroscopy, thermogravimetric analysis, differential scanning calorimetry, and scanning electron microscopy. The maximum adsorption of RB is 193.05 mg/g and could be well described by the Langmuir isotherm and pseudo kinetic order model. Desorption of 93% of RB occurred at pH 4.01, a temperature of 60°C, and within 10 hr. With an increase in the amount of acrylic acid, there was an increase in the pH sensitivity of the hydrogel. With the change in pH value, the volume of the hydrogel varied. Due to its easy adsorption-desorption process and regeneration ability, it might lead to wide utilization in the domain of wastewater management.

Utilization of thermal and pH dual-responsive cellulose microfilaments/poly (N-Isopropylacrylamide-co-acrylic acid) spheres (MPNAA) for the elimination of dye (methylene blue (MB) and methyl violet (MV)), in solitary and dual arrangements, was examined by Yuan Lia *et al.* [65]. MPNAA displayed utmost adsorption ability of 497.5 and 840.3 mg g^{-1} towards MB and MV, respectively, in single systems, and high adsorption capacity was maintained in binary systems with the favorable adsorption of MV at pH 9, 20 ppm of initial concentration of MB, temperature at 30 °C, and the adsorbent dosage at 1000 mg/L. As the adsorbents under neutral or considerably alkaline medium are negatively charged, at higher pH a strong ionic interaction is established between the smart polymers and the dyes. The desorption efficacy after the fifth time of reprocessing and at a temperature of 60°C appeared as 87.12% for MB and 85.79% for MV.

Flocculation is one of the widely used methods for the treatment of water contaminated with heavy metal ions and antibiotics. Zhen Yang and associates also testified to the synthesis and application of a novel pH- and temperature-responsive flocculant, carboxymethyl chitosan-graft-poly(N-isoproyl acrylamide-co-diallyl dimethyl ammonium chloride) (denoted as CND) for the removal of copper(II) and tetracycline (TC) from livestock wastewater [66]. The optimal flocculation performance of CND would take place within the pH range of 6.5–8.0 and at two temperatures, 25 and 45°C. On rising of pH to 7, strong charge attraction took place between the positively charged flocculant and negatively charged contaminants; however, the net charge of CND became near zero when pH was 8, leading to decreased flocculation efficiency. CND enhanced the "mutual-reinforcement" effect between copper(II) and TC in flocculation, which increased the removal rate. Zhen Yang *et al.* [67] developed a pH- and temperature-responsive magnetic adsorbent [poly(N-isopropyl acrylamide) grafted chitosan/Fe_3O_4 site particles, CN-MCP], for the removal of the endocrine-disrupting chemical nonylphenol (NP). CN-MCP showed a relatively high adsorption capacity of 123 mg/g for nonylphenol under corresponding optimal conditions at pH=9 and 20°C. The presence of other ions had very little effect on the rate of adsorption of NP. The regeneration of the adsorbent took place at pH 5 and 40°C temperature and could be recycled up to 9 times with similar efficiency.

CONCLUSION

This book chapter has encapsulated the different categories of stimuli-responsive materials that have been employed for water purification. These materials are unique, and on the basis of the nature of the stimulants that exist inside or outside, they display changeable qualities and show great potential performance in wastewater treatment. Along with their potential, these materials have some limitations. *e.g.*, in the case of materials that are sensitive to pH, shifting the value of pH involves a large quantity of compounds, whereas splitting of solutes occurs with the variation in the dimension of the pore of the matters. The array of temperatures for heat-responsive substances is not very broad, *e.g.*, 10°C-15°C. The establishment of the exterior field for both electric and magnetic field-sensitive polymers is cost-effective. Excluding these limitations, stimuli-responsive polymers overcome the problems associated with water treatment. The selectivity, competency, repeatable nature, and high ability to spot harmful contaminants make the smart materials an alternative means for water purification. In the future, biodegradable polymer combinations with smart materials will be capable of portraying a significant function in wastewater management.

REFERENCES

[1] Stuart, M.A.C.; Huck, W.T.S.; Genzer, J.; Müller, M.; Ober, C.; Stamm, M.; Sukhorukov, G.B.; Szleifer, I.; Tsukruk, V.V.; Urban, M.; Winnik, F.; Zauscher, S.; Luzinov, I.; Minko, S. Emerging applications of stimuli-responsive polymer materials. *Nat. Mater.*, **2010**, *9*(2), 101-113. [http://dx.doi.org/10.1038/nmat2614] [PMID: 20094081]

[2] Lewis, S.R.; Datta, S.; Gui, M.; Coker, E.L.; Huggins, F.E.; Daunert, S.; Bachas, L.; Bhattacharyya, D. Reactive nanostructured membranes for water purification. *Proc. Natl. Acad. Sci. USA,* **2011**, *108*(21), 8577-8582. [http://dx.doi.org/10.1073/pnas.1101144108] [PMID: 21606340]

[3] Hu, J.; Liu, S. Responsive polymers for detection and sensing applications. *Current Status and Future Developments Macromolecules,* **2010**, *43*, 8315-8330.

[4] Mura, S.; Nicolas, J.; Couvreur, P. Stimuli-responsive nanocarriers for drug delivery. *Nat. Mater.,* **2013**, *12*(11), 991-1003. [http://dx.doi.org/10.1038/nmat3776] [PMID: 24150417]

[5] Kumar, S.; Dory, Y. L.; Lepage, M.; Zhao, Y. Surface-grafted stimuli-responsive block copolymer brushes for the thermo-, photo- and pH-sensitive release of dye molecules macromolecules **2011**, *44*, 7385-7393.

[6] Dutta, K.; Mahale, R. Y.; Arulkashmir, A.; Krishnamoorthy, K. *Reversible Assembly and Disassembly of Micelles by a Polymer That Switches between Hydrophilic and Hydrophobic Wettings Langmuir,* **2012**, *28*, 10097-10104.

[7] Nistor, V.; Cannell, J.; Gregory, J.; Yeghiazarian, L. Stimuli-responsive cylindrical hydrogels mimic intestinal peristalsis to propel a solid object. *Soft Matter,* **2016**, *12*(15), 3582-3588. [http://dx.doi.org/10.1039/C5SM02553B] [PMID: 26971454]

[8] Roy, D.; Cambre, J.N.; Sumerlin, B.S. Future perspectives and recent advances in stimuli-responsive materials. *Prog. Polym. Sci.,* **2010**, *35*(1-2), 278-301. [http://dx.doi.org/10.1016/j.progpolymsci.2009.10.008]

[9] Xu, T.; Huang, C. Electrodialysis-based separation technologies: A critical review. *AIChE J.,* **2008**, *54*(12), 3147-3159. [http://dx.doi.org/10.1002/aic.11643]

[10] Babel, S.; Kurniawan, T.A. Low-cost adsorbents for heavy metals uptake from contaminated water: a review. *J. Hazard. Mater.,* **2003**, *97*(1-3), 219-243. [http://dx.doi.org/10.1016/S0304-3894(02)00263-7] [PMID: 12573840]

[11] Fu, F.; Wang, Q. Removal of heavy metal ions from wastewaters: A review. *J. Environ. Manage.,* **2011**, *92*(3), 407-418. [http://dx.doi.org/10.1016/j.jenvman.2010.11.011] [PMID: 21138785]

[12] Dutta, K.; De, S. Aromatic conjugated polymers for removal of heavy metal ions from wastewater: a short review. *Environ. Sci. Water Res. Technol.,* **2017**, *3*(5), 793-805. [http://dx.doi.org/10.1039/C7EW00154A]

[13] Lee, A.; Elam, J.W.; Darling, S.B. Membrane materials for water purification: design, development, and application. *Environ. Sci. Water Res. Technol.,* **2016**, *2*(1), 17-42. [http://dx.doi.org/10.1039/C5EW00159E]

[14] Cai, Y.; Shen, W.; Loo, S.L.; Krantz, W.B.; Wang, R.; Fane, A.G.; Hu, X. Towards temperature driven forward osmosis desalination using Semi-IPN hydrogels as reversible draw agents. *Water Res.,* **2013**, *47*(11), 3773-3781. [http://dx.doi.org/10.1016/j.watres.2013.04.034] [PMID: 23726714]

[15] Liu, Z.; Luo, F.; Ju, X.J.; Xie, R.; Sun, Y.M.; Wang, W.; Chu, L.Y. Gating membranes for water treatment: detection and removal of trace Pb2+ ions based on molecular recognition and polymer phase transition. *J. Mater. Chem. A Mater. Energy Sustain.,* **2013**, *1*(34), 9659-9671. [http://dx.doi.org/10.1039/c3ta12006f]

[16] Abousalman-Rezvani, Z.; Eskandari, P.; Roghani-Mamaqani, H.; Salami-Kalajahi, M. Synthesis of coumarin-containing multi-responsive CNC-grafted and free copolymers with application in nitrate ion removal from aqueous solutions. *Carbohydr. Polym.,* **2019**, *225*, 115247. [http://dx.doi.org/10.1016/j.carbpol.2019.115247] [PMID: 31521262]

[17] Abousalman-Rezvani, Z.; Eskandari, P.; Roghani-Mamaqani, H.; Mardani, H.; Salami-Kalajahi, M. Grafting light-, temperature, and CO2-responsive copolymers from cellulose nanocrystals by atom transfer radical polymerization for adsorption of nitrate ions. *Polymer (Guildf.),* **2019**, *182*, 121830. [http://dx.doi.org/10.1016/j.polymer.2019.121830]

[18] Qin, Y.; Wang, L.; Zhao, C.; Chen, D.; Ma, Y.; Yang, W. Ammonium-functionalized hollow polymer particles as a ph-responsive adsorbent for selective removal of acid dye. *ACS Appl. Mater. Interfaces,* **2016**, *8*(26), 16690-16698. [http://dx.doi.org/10.1021/acsami.6b04199] [PMID: 27302068]

[19] Xu, B.; Zheng, H.; Zhou, H.; Wang, Y.; Luo, K.; Zhao, C.; Peng, Y.; Zheng, X. Adsorptive removal of anionic dyes by chitosan-based magnetic microspheres with pH-responsive properties. *J. Mol. Liq.,* **2018**, *256*, 424-432. [http://dx.doi.org/10.1016/j.molliq.2018.02.061]

[20] Ifra, ; Saha, S. Fabrication of topologically anisotropic microparticles and their surface modification with pH responsive polymer brush. *Mater. Sci. Eng. C,* **2019**, *104*, 109894. [http://dx.doi.org/10.1016/j.msec.2019.109894] [PMID: 31499968]

[21] Safavi-Mirmahalleh, S.A.; Salami-Kalajahi, M.; Roghani-Mamaqani, H. Adsorption kinetics of methyl orange from water by pH-sensitive poly(2-(dimethylamino)ethyl methacrylate)/nanocrystalline cellulose hydrogels. *Environ. Sci. Pollut. Res. Int.,* **2020**, *27*(22), 28091-28103. [http://dx.doi.org/10.1007/s11356-020-09127-y] [PMID: 32405949]

[22] Duan, Y.; Song, Y.; Zhou, L. Facile synthesis of polyamidoamine dendrimer gel with multiple amine groups as a super adsorbent for highly efficient and selective removal of anionic dyes. *J. Colloid Interface Sci.,* **2019**, *546*, 351-360. [http://dx.doi.org/10.1016/j.jcis.2019.03.073] [PMID: 30928728]

[23] Abousalman-Rezvani, Z., Eskandari, P., Roghani-Mamaqani, H., Mardani, H. and Salami-Kalajahi, M., Grafting light-, temperature, and CO2-responsive copolymers from cellulose nanocrystals by atom transfer radical polymerization for adsorption of nitrate ions. *Polymer,* **2019**, *182*, 121830.

[24] Yang, L.; Zhan, Y.; Yu, R.; Lan, J.; Shang, J.; Dou, B.; Liu, H.; Zou, R.; Lin, S. Facile and scalable fabrication of antibacterial CO_2 -responsive cotton for ultrafast and controllable removal of anionic dyes. *ACS Appl. Mater. Interfaces,* **2021**, *13*(2), 2694-2709. [http://dx.doi.org/10.1021/acsami.0c19750]

[25] Kaşgöz, H.; Durmuş, A.; Kaşgöz, A. Enhanced swelling and adsorption properties of AAm-AMPSNa/clay hydrogel nanocomposites for heavy metal ion removal. *Polym. Adv. Technol.,* **2008**, *19*(3), 213-220. [http://dx.doi.org/10.1002/pat.999]

[26] Atta, A.M.; Ismail, H.S.; Elsaaed, A.M. Application of anionic acrylamide-based hydrogels in the removal of heavy metals from waste water. *J. Appl. Polym. Sci.,* **2012**, *123*(4), 2500-2510. [http://dx.doi.org/10.1002/app.34798]

[27] Ramírez, E.; Burillo, S.G.; Barrera-Díaz, C.; Roa, G.; Bilyeu, B. Use of pH-sensitive polymer hydrogels in lead removal from aqueous solution. *J. Hazard. Mater.,* **2011**, *192*(2), 432-439. [http://dx.doi.org/10.1016/j.jhazmat.2011.04.109] [PMID: 21705138]

[28] Li, Z.; Wang, Y.; Wu, N.; Chen, Q.; Wu, K. Removal of heavy metal ions from wastewater by a novel HEA/AMPS copolymer hydrogel: preparation, characterization, and mechanism. *Environ. Sci. Pollut. Res. Int.,* **2013**, *20*(3), 1511-1525. [http://dx.doi.org/10.1007/s11356-012-0973-2] [PMID: 22614052]

[29] Hou, N.; Wang, R.; Wang, F.; Bai, J.; Jiao, T.; Bai, Z.; Zhang, L.; Zhou, J.; Peng, Q. Self-assembled hydrogels constructed *via* host-guest polymers with highly efficient dye removal capability for wastewater treatment. *Colloids Surf. A Physicochem. Eng. Asp.,* **2019**, *579*, 123670. [http://dx.doi.org/10.1016/j.colsurfa.2019.123670]

[30] Singh, R.; Pal, D.; Mathur, A.; Singh, A.; Krishnan, M.A.; Chattopadhyay, S. An efficient pH sensitive hydrogel, with biocompatibility and high reusability for removal of methylene blue dye from aqueous solution. *React. Funct. Polym.,* **2019**, *144*, 104346. [http://dx.doi.org/10.1016/j.reactfunctpolym.2019.104346]

[31] Bidarakatte Krishnappa, P.; Badalamoole, V. Karaya gum-graft-poly(2-(dimethylamino)ethyl methacrylate) gel: An efficient adsorbent for removal of ionic dyes from water. *Int. J. Biol. Macromol.,* **2019**, *122*, 997-1007. [http://dx.doi.org/10.1016/j.ijbiomac.2018.09.038] [PMID: 30201563]

[32] Ali, S.A.; Kazi, I.W.; Ullah, N. New chelating ion-exchange resin synthesized *via* the cyclopolymerization protocol and its uptake performance for metal ion removal ind. Eng. *Chem. Res.,* **2015**, *54*, 9689-9698.

[33] Sánchez, J.; Espinosa, C.; Pooch, F.; Tenhu, H.; Pizarro, G.C.; Oyarzún, D.P. Poly(N,N-dimethylaminoethyl methacrylate) for removing chromium (VI) through polymer-enhanced ultrafiltration technique. *React. Funct. Polym.,* **2018**, *127*, 67-73. [http://dx.doi.org/10.1016/j.reactfunctpolym.2018.04.002]

[34] Liu, B.; Chen, X.; Zheng, H.; Wang, Y.; Sun, Y.; Zhao, C.; Zhang, S. Rapid and efficient removal of heavy metal and cationic dye by carboxylate-rich magnetic chitosan flocculants: Role of ionic groups. *Carbohydr. Polym.,* **2018**, *181*, 327-336. [http://dx.doi.org/10.1016/j.carbpol.2017.10.089] [PMID: 29253979]

[35] Ali, S.A.; Rachman, I.B.; Saleh, T.A. Simultaneous trapping of Cr(III) and organic dyes by a pH-responsive resin containing zwitterionic aminomethylphosphonate ligands and hydrophobic pendants. *Chem. Eng. J.,* **2017**, *330*, 663-674. [http://dx.doi.org/10.1016/j.cej.2017.08.003]

[36] Tang, J.; Song, Y.; Zhao, F.; Spinney, S.; da Silva Bernardes, J.; Tam, K.C. Compressible cellulose nanofibril (CNF) based aerogels produced *via* a bio-inspired strategy for heavy metal ion and dye removal. *Carbohydr. Polym.,* **2019**, *208*, 404-412. [http://dx.doi.org/10.1016/j.carbpol.2018.12.079] [PMID: 30658817]

[37] Tran, N.B.; Kim, J.Y.; Kim, Y.C.; Kim, Y.J.; Kim, J.H. CO_2 -responsive swelling behavior and metal-ion adsorption properties in novel histamine-conjugated polyaspartamide hydrogel. *J. Appl. Polym. Sci.,* **2016**, *133*(16), app.43305. [http://dx.doi.org/10.1002/app.43305]

[38] Chen, J.J.; Ahmad, A.L.; Ooi, B.S. Poly(N-isopropylacrylamide-co-acrylic acid) hydrogels for copper ion adsorption: Equilibrium isotherms, kinetic and thermodynamic studies. *J. Environ. Chem. Eng.,* **2013**, *1*(3), 339-348. [http://dx.doi.org/10.1016/j.jece.2013.05.012]

[39] Bai, Y.; Liang, Y.N.; Hu, X. An eco-friendly approach for heavy metal adsorbent regeneration using CO_2-responsive molecular octopus. *Chemosphere,* **2017**, *185*, 1157-1163. [http://dx.doi.org/10.1016/j.chemosphere.2017.07.086] [PMID: 28764136]

[40] Fan, S.; Chen, J.; Fan, C.; Chen, G.; Liu, S.; Zhou, H.; Liu, R.; Zhang, Y.; Hu, H.; Huang, Z.; Qin, Y.; Liang, J. Fabrication of a CO_2-responsive chitosan aerogel as an effective adsorbent for the adsorption and desorption of heavy metal ions. *J. Hazard. Mater.,* **2021**, *416*, 126225. [http://dx.doi.org/10.1016/j.jhazmat.2021.126225] [PMID: 34492979]

[41] Ito, T.; Hioki, T.; Yamaguchi, T.; Shinbo, T.; Nakao, S.; Kimura, S. Development of a molecular recognition ion gating membrane and estimation of its pore size control. *J. Am. Chem. Soc.,* **2002**, *124*(26), 7840-7846. [http://dx.doi.org/10.1021/ja012648x] [PMID: 12083938]

[42] Xu, F.J.; Li, J.; Yuan, S.J.; Zhang, Z.X.; Kang, E.T.; Neoh, K.G. Thermo-responsive porous membranes of controllable porous morphology from triblock copolymers of polycaprolactone and poly(N-isopropylacrylamide) prepared by atom transfer radical polymerization. *Biomacromolecules,* **2008**, *9*(1), 331-339. [http://dx.doi.org/10.1021/bm7008922] [PMID: 18067254]

[43] Rao, G.V.R.; Krug, M.E.; Balamurugan, S.; Xu, H.; Xu, Q.; López, G.P. Synthesis and Characterization of Silica−Poly(N-isopropylacrylamide). *Hybrid Membranes: Switchable Molecular Filters Chem. Mater.,* **2002**, *14*, 5075-5080.

[44] Fu, Q.; Rao, G. V. R.; Ward, T. L.; Lu, Y.; Lopez, G. P. *Thermoresponsive Transport through Ordered Mesoporous Silica/PNIPAAm Copolymer Membranes and Microspheres Langmuir,* **2007**, *23*, 170-174.

[45] Shevate, R.; Karunakaran, M.; Kumar, M.; Peinemann, K.V. Polyanionic pH-responsive polystyrene-b-poly(4-vinyl pyridine-N-oxide) isoporous membranes. *J. Membr. Sci.,* **2016**, *501*, 161-168. [http://dx.doi.org/10.1016/j.memsci.2015.11.038]

[46] Himstedt, H.H.; Marshall, K.M.; Wickramasinghe, S.R. pH-responsive nanofiltration membranes by surface modification. *J. Membr. Sci.,* **2011**, *366*(1-2), 373-381. [http://dx.doi.org/10.1016/j.memsci.2010.10.027]

[47] Lee, D.; Nolte, A.J.; Kunz, A.L.; Rubner, M.F.; Cohen, R.E. pH-induced hysteretic gating of track-etched polycarbonate membranes: swelling/deswelling behavior of polyelectrolyte multilayers in confined geometry. *J. Am. Chem. Soc.,* **2006**, *128*(26), 8521-8529. [http://dx.doi.org/10.1021/ja0608803] [PMID: 16802818]

[48] Lu, T.; Xiang, T.; Huang, X.L.; Li, C.; Zhao, W.F.; Zhang, Q.; Zhao, C.S. Post-crosslinking towards stimuli-responsive sodium alginate beads for the removal of dye and heavy metals. *Carbohydr. Polym.,* **2015**, *133*, 587-595. [http://dx.doi.org/10.1016/j.carbpol.2015.07.048] [PMID: 26344317]

[49] Kostal, J.; Mulchandani, A.; Gropp, K.E.; Chen, W. A temperature responsive biopolymer for mercury remediation. *Environ. Sci. Technol.,* **2003**, *37*(19), 4457-4462. [http://dx.doi.org/10.1021/es034210y] [PMID: 14572100]

[50] Yamashita, K.; Nishimura, T.; Nango, M. Preparation of IPN-type stimuli-Responsive heavy-Meta--Ion adsorbent gel. *Polym. Adv. Technol.,* **2003**, *14*(3-5), 189-194. [http://dx.doi.org/10.1002/pat.290]

[51] Yamashita, K.; Nishimura, T.; Ohashi, K.; Ohkouchi, H.; Nango, M. Two-step imprinting procedure of inter-penetrating polymer network-type stimuli-responsive hydrogel-adsorbents. *Polym. J.,* **2003**, *35*(7), 545-550. [http://dx.doi.org/10.1295/polymj.35.545]

[52] Parasuraman, D.; Serpe, M.J. Poly (N-isopropylacrylamide) microgel-based assemblies for organic dye removal from water. *ACS Appl. Mater. Interfaces,* **2011**, *3*(12), 4714-4721.

[http://dx.doi.org/10.1021/am201132x] [PMID: 22053867]

[53] Fries, K.H.; Driskell, J.D.; Sheppard, G.R.; Locklin, J. Fabrication of spiropyran-containing thin film sensors used for the simultaneous identification of multiple metal ions. *Langmuir,* **2011**, *27*(19), 12253-12260.
[http://dx.doi.org/10.1021/la202344w] [PMID: 21877693]

[54] Genovese, M.E.; Caputo, G.; Nanni, G.; Setti, C.; Bustreo, M.; Perotto, G.; Athanassiou, A.; Fragouli, D. Light responsive silk nanofibers: An optochemical platform for environmental applications. *ACS Appl. Mater. Interfaces,* **2017**, *9*(46), 40707-40715.
[http://dx.doi.org/10.1021/acsami.7b13372] [PMID: 29120601]

[55] He, S.; Xiao, L.; Marin, L.; Bai, Y.; Cheng, X. Fully-water-soluble BODIPY containing fluorescent polymers prepared by RAFT method for the detection of Fe3+ ions. *Eur. Polym. J.,* **2021**, *150*, 110428.
[http://dx.doi.org/10.1016/j.eurpolymj.2021.110428]

[56] Ozay, O.; Ekici, S.; Baran, Y.; Kubilay, S.; Aktas, N.; Sahiner, N. Utilization of magnetic hydrogels in the separation of toxic metal ions from aqueous environments. *Desalination,* **2010**, *260*(1-3), 57-64.
[http://dx.doi.org/10.1016/j.desal.2010.04.067]

[57] Si, Y.; Ren, T.; Li, Y.; Ding, B.; Yu, J. Fabrication of magnetic polybenzoxazine-based carbon nanofibers with Fe3O4 inclusions with a hierarchical porous structure for water treatment. *Carbon,* **2012**, *50*(14), 5176-5185.
[http://dx.doi.org/10.1016/j.carbon.2012.06.059]

[58] Wang, Z.; Feng, Y.; Hao, X.; Huang, W.; Feng, X. A novel potential-responsive ion exchange film system for heavy metal removal. *J. Mater. Chem. A Mater. Energy Sustain.,* **2014**, *2*(26), 10263-10272.
[http://dx.doi.org/10.1039/C4TA00782D]

[59] Wang, Z.; Guo, S.; Wu, Z.; Fan, H.; Guan, G.; Hao, X. A smart potential-responsive ion exchange nanomaterial with superparamagnetism for cesium ion separation and recovery. *Separ. Purif. Tech.,* **2017**, *187*, 199-206.
[http://dx.doi.org/10.1016/j.seppur.2017.06.033]

[60] Shirsath, S.R.; Hage, A.P.; Zhou, M.; Sonawane, S.H.; Ashokkumar, M. Ultrasound assisted preparation of nanoclay Bentonite-FeCo nanocomposite hybrid hydrogel: A potential responsive sorbent for removal of organic pollutant from water. *Desalination,* **2011**, *281*, 429-437.
[http://dx.doi.org/10.1016/j.desal.2011.08.031]

[61] Venkata Mohan, S.; Chandrasekhar Rao, N.; Krishna Prasad, K.; Karthikeyan, J.; Karthikeyan, J. Treatment of simulated Reactive Yellow 22 (Azo) dye effluents using Spirogyra species. *Waste Manag.,* **2002**, *22*(6), 575-582.
[http://dx.doi.org/10.1016/S0956-053X(02)00030-2] [PMID: 12214968]

[62] Özkahraman, B.; Acar, I.; Emik, S. Removal of cationic dyes from aqueous solutions with poly (N-isopropylacrylamide-co-itaconic acid) hydrogels. *Polym. Bull.,* **2011**, *66*(4), 551-570.
[http://dx.doi.org/10.1007/s00289-010-0371-1]

[63] Karthika, J.S.; Vishalakshi, B. Novel stimuli responsive gellan gum-graft-poly(DMAEMA) hydrogel as adsorbent for anionic dye. *Int. J. Biol. Macromol.,* **2015**, *81*, 648-655.
[http://dx.doi.org/10.1016/j.ijbiomac.2015.08.064] [PMID: 26325677]

[64] Yao, G.; Li, S.; Xu, J.; Liu, H. Dual-responsive graphene oxide/poly(NIPAM- *co* -AA) hydrogel as an adsorbent for rhodamine b and imidacloprid. *J. Chem. Eng. Data,* **2019**, *64*(9), 4054-4065.
[http://dx.doi.org/10.1021/acs.jced.9b00408]

[65] Li, Y.; Xiao, H.; Pan, Y.; Zhang, M.; Jin, Y. Thermal and pH dual-responsive cellulose microfilament spheres for dye removal in single and binary systems. *J. Hazard. Mater.,* **2019**, *377*, 88-97.
[http://dx.doi.org/10.1016/j.jhazmat.2019.05.033] [PMID: 31153117]

[66] Yang, Z.; Jia, S.; Zhuo, N.; Yang, W.; Wang, Y. Flocculation of copper(II) and tetracycline from water using a novel pH- and temperature-responsive flocculants. *Chemosphere,* **2015**, *141*, 112-119. [http://dx.doi.org/10.1016/j.chemosphere.2015.06.050] [PMID: 26162528]

[67] Zhen, Y.; Ning, Z.; Shaopeng, Z.; Yayi, D.; Xuntong, Z.; Jiachun, S.; Weiben, Y.; Yuping, W.; Jianqiang, C. A pH- and temperature-responsive magnetic composite adsorbent for targeted removal of nonylphenol. *ACS Appl. Mater. Interfaces,* **2015**, *7*(44), 24446-24457. [http://dx.doi.org/10.1021/acsami.5b08709] [PMID: 26492983]

CHAPTER 7

Stimuli-Responsive Polymeric Nanocomposites and their Applications

Geeti Kaberi Dutta[1,*] and **Nobomi Borah**[2]

[1] *Department of Chemistry, Bahona College, Bahona, Jorhat-785101, Assam, India*

[2] *Department of Chemical Sciences, Tezpur University, Tezpur-784028, Assam, India*

Abstract: As technology globally embraces "smart" features, the polymer industry has acquired some advanced creations, including the development of polymer nanocomposites (PNCs) characterized by smart or intelligent properties. These smart features originate from the ability of these materials to respond to external cues such as pH, temperature, stress, light, electricity, magnetism, chemical environments, *etc*., facilitating their end use as flexible electronic or motion sensors, shape recognizable polymers, and self-healable materials. The sensitivity of the materials can be hierarchical within the polymer matrix or can be adventitiously introduced by the nanomaterials resulting in a combined system that reflects some additional characteristics. This chapter aims to highlight the various aspects of stimuli-responsive smart PNCs; including a brief overview of the fabrication methods of PNCs using different polymer matrices such as polyurethanes, polyesters, epoxy resins, hydrogels, *etc*., and various advanced metal- or carbon-based nanomaterials. The authentication methods as well as various sophisticated attributes of these materials are highlighted in the chapter. In turn, the numerous applications of these PNCs in real-world scenarios are also discussed briefly. Thus, the chapter represents an attempt to illustrate the fabrication methods of PNCs and their frontline applications that help in understanding their future possibilities in the area of material sciences.

Keywords: Drug delivery, Electro responsive, Magneto responsive, Mechanical, Nanostructure, Optical, pH-responsive, Photo-responsive, Polymer nanocomposites, Sensors, Solubility, Thermal stability, Thermo-responsive.

INTRODUCTION

Nature offers a remarkable array of examples showcasing stimuli-responsive behaviours in living organisms. Sunflowers exhibit heliotropism where the flower heads track the movement of the sun, chameleons change color in accordance to their environment, *Mimosa pudica* commonly known as the "touch-me-not" plant

* **Corresponding author Geeti Kaberi Dutta:** Department of Chemistry, Bahona College, Bahona, Jorhat-785101, Assam, India; E-mail: geetikaberi404@gmail.com

exhibits rapid leaf movement in response to touch, *Codariocalyx motorius* exhibits leaflet movement in response to various stimuli (light, touch, or sound); and many more. These examples highlight the remarkable adaptability and responsiveness of living organisms to their surroundings [1, 2]. Scientists draw inspiration from such natural phenomena to develop stimuli-responsive materials and technologies with applications ranging from robotics and biomimetics to smart materials and environmental sensing. Mimicking these natural mechanisms in synthetic materials not only led to significant advancements in the field of materials science but also opened up new avenues for innovation in various fields [3].

Polymers play a crucial role in material science and their characteristics showed great potential to yield smart materials. Thus, polymers have become one of the main research directions among stimuli-responsive materials since they have the ability to give remarkable responses to small changes. Living organisms showing stimuli-responsive behaviour are made up of natural polymers such as proteins, polysaccharides, and nucleic acids, and to mimic these natural mechanisms of stimuli responsiveness, synthetic polymers have been developed with similar functionalities [4]. Synthetic polymers can be classified into different categories based on their chemical properties, structure, and behaviour; such as thermoplastics, thermosetting polymers, elastomers, biodegradable polymers, smart polymers, *etc*. Continued research and innovation in polymer science and engineering hold the potential to further expand the range of applications of synthetic polymers as stimuli responsive polymers (SRPs). SRPs are also called stimuli-sensitive, smart, environmentally sensitive, and adaptive polymers [5, 6]. These polymers typically consist of repeating units with functional groups that can undergo reversible changes in conformation ensuing change in their physical or chemical properties in response to external stimuli. The most common stimuli that trigger responses in these polymers include temperature [7], light [8], pH [9], electric field [10], magnetic field [11], and oxidation-reduction [12, 13]. Different responses can be observed by the SRPs depending on the specific properties being altered and the above-mentioned stimuli triggering these changes (Fig. **1**). SRPs change the overall polymer chain dimensions or size by inducing the expansion or contraction of polymer chains. For example, temperature-responsive polymers may undergo coil-to-globule transitions, where the polymer chains either collapse or swell in response to changes in temperature [14]. Some SRPs exhibit changes in their secondary structure in response to stimuli such as temperature, pH, or solvent composition. For instance, temperature-sensitive polymers may undergo transitions between ordered and disordered states, affecting their mechanical properties or interactions with other molecules [15]. Further, SRPs can undergo changes in solubility or dispersibility in solution upon exposure to specific stimuli, which may involve transitions between soluble and insoluble states, leading to phase separation or precipitation of the polymer. SRPs may also undergo changes in the degree of intermolecular association or aggregation in response to stimuli. For example, pH-sensitive polymers may form aggregates or micelles in response to changes in pH, affecting their behavior in solution. The reasons for the above-mentioned responses often involve the formation or disruption of secondary forces such as hydrogen bonding, hydrophobic interactions, or electrostatic interactions; simple acid-base reactions of moieties branched to the polymer backbone, conformational changes, crosslinking/degradation reactions, or phase transitions leading to the formation of new structures or morphologies [16, 17].

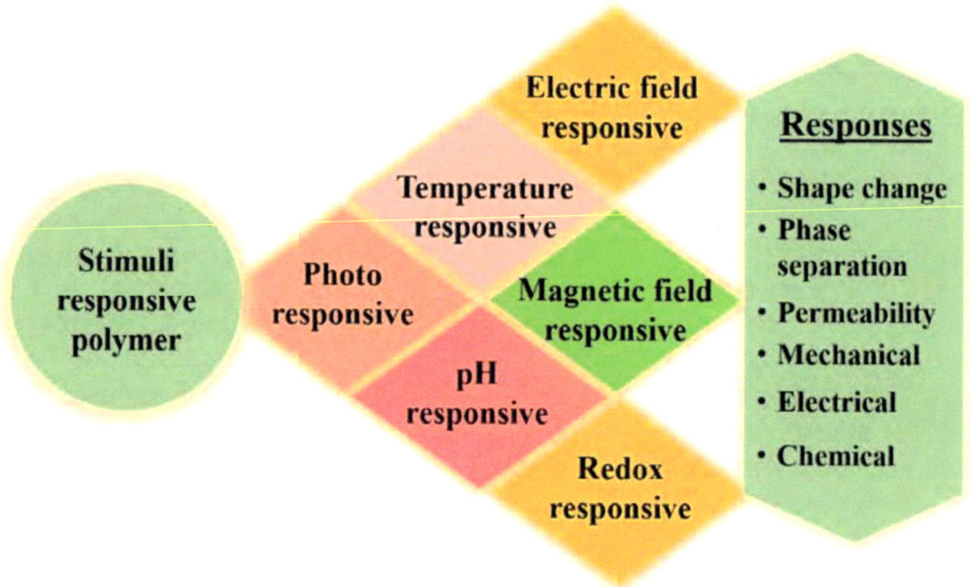

Fig. (1). Potential stimuli and responses of SRPs.

The stimuli sensitivity can be introduced into a polymer or stimuli responsiveness of SRPs can be enhanced by modifying the pristine polymer backbone chemically via the incorporation of multiple responsive units such as small polymer fragments or functional groups with high stimuli sensitivity. Further, nanomaterials can also be incorporated into polymer matrix leading to the formation of polymer nanocomposites (PNCs) either to add stimuli sensitivity into the polymer or to enhance stimuli responsiveness of the SRPs along with other properties making them suitable for practical use [18]. The high specific surface area, reinforcing effects and inherent functionalities of the nanomaterials help in introducing and enhancing stimuli responsiveness in the pristine polymers and SRPs, respectively along with enhancement of the overall performance. The addition of such multiple responsive units allows synergistic effects in the polymer chain in response to stimuli, where relatively minor changes in chemical structure can result in dramatic changes in macroscopic material properties. The stimuli responsiveness in the PNCs can be tuned according to the requirement by varying the amount of the incorporated nanomaterials [19]. Furthermore, in addition to single stimuli response, modern SRPs or stimuli-responsive polymer nanocomposites (SRPNCs) can be designed to sense and respond to multiple stimulating factors simultaneously through special molecular design strategies [20]. Thus, the stimuli responsiveness and its tunability along with the other properties make SRPNCs a promising avenue in materials science, offering solutions to address the limitations of traditional pristine polymers and enabling the creation of innovative materials with enhanced performance and functionality for various applications [21 - 24]. This chapter reviews the fabrication strategies of SRPNCs that are sensitive to various stimuli like heat, light, electrical field, magnetic field, *etc*. Further, wide applications of SRPNCs are also summarized in this chapter.

STIMULI-RESPONSIVE POLYMER NANOCOMPOSITES AND THEIR FABRICATION TECHNIQUES

Although polymers hold a crucial place in the field of material science because of their exceptional properties, however, the use of bare polymers is sometimes limited in the practical field because of their low mechanical properties, low conductivity, low resistance towards the harsh environment, low or no stimuli sensitivity, heavy weight, *etc*. One alternative to overcome such poor properties is to incorporate nanomaterials into the polymer matrix to form PNCs. The formation of PNCs has indeed revolutionized material science, offering a pathway to enhance and tailor properties that were previously limited in polymers alone. Nanomaterials in PNCs not only help in improving the existing properties of the polymers but also contribute to the generation of sensitivity towards different stimuli making them more versatile and functional for various applications [18, 25]. Although nanomaterials can be classified according to their organic or inorganic origin, and different dimensions; they also can be divided according to their ability to respond to stimuli as inert or active nanofillers. Inert nanofillers are those which does not respond to any stimuli and hence they primarily serve as mechanical reinforcements for polymers along with the enhancement of other properties. Active nanofillers possess stimulus-responsive properties, enabling them to respond to external stimuli and hence they not only reinforce the polymer matrix but also imbue the nanocomposite with the ability to react and adapt to different environmental conditions or stimuli [26, 27]. For example, the incorporation of nanomaterials that react with light, temperature, pH, magnetic field, *etc*., allows the PNCs to undergo structural changes followed by a change in their properties in accordance with these stimuli, making them suitable for applications such as optical switches, sensors, thermal actuators, drug delivery systems, diagnostics, shape-memory materials, magnetic sensors, and self-healing materials [15 - 18, 21 - 24].

The fabrication of SRPNCs predominantly employs active nanomaterials due to their ability to impart responsive properties to the material. By carefully selecting and integrating these active nanomaterials into polymer matrices, researchers can design nanocomposites with tailored responses to specific stimuli. The fabrication of SRPNCs is indeed a complex process that requires careful consideration of various factors, including the characteristics of the polymer matrix, the nanofiller, and their interactions. Different processing techniques have been developed to address these challenges and achieve uniform dispersion of nanofillers within the polymer matrix. Some of the most commonly used techniques include *in situ* polymerization, melt compounding, and solution mixing. In *in situ* polymerization, the nanomaterials are mixed with the monomer solution at the beginning of the polymerization reaction and as the polymerization progresses, the nanomaterials get dispersed within the polymer matrix [28]. In melt compounding, the polymer is first melted at elevated temperatures minimizing the need for solvent in the polymerization process and the nanomaterials are mixed with that molten polymer. However, the high temperatures required for this method can sometimes affect the properties of the nanomaterials or the polymer matrix [29]. Further, in solution mixing, the nanomaterial is first dispersed in a solvent which is later mixed with the polymer solution prepared by dissolving the polymer in a suitable solvent. After homogenization, the solvent is evaporated to obtain nanomaterials dispersed in the polymer matrix. Although, this method is relatively easy to perform and requires lower temperatures, however, the selection of the solvent is critical to ensure proper dispersion of nanomaterials and dissolution of the polymer [30]. Each of these synthesis methods has its own advantages and limitations, and the choice of method depends on factors such as the nature of the polymer and nanomaterials, desired properties of the nanocomposite, and processing conditions [31].

DIFFERENT STIMULI-RESPONSIVE POLYMER NANOCOMPOSITES

A SRPNC material is typically engineered to exhibit spontaneous responses triggered by external stimuli through alterations in its structure and some of the most common stimuli responsiveness of PNCs are discussed below:

Photo-responsive Polymer Nanocomposites

Photo-responsive PNCs are materials that can undergo changes in their properties upon exposure to light and these changes can include alterations in colour, shape, mechanical properties, optical properties, solubility, wettability, conductivity, *etc*. These materials can be obtained either by incorporating photosensitive moieties or chromophores in the polymeric chain during their polymerization or by adding chromophoric nanomaterials to the PNC matrix [32]. The presence of chromophoric nanomaterials in PNCs can enhance the photo responsiveness of the polymer matrix with their plasmonic or semiconductor properties, which can amplify light absorption capacity and facilitate charge transfer processes. Depending on the specific chromophores used and their photochemical properties, the resulting photo-responsive PNCs can exhibit either reversible or irreversible responses to light stimuli. Reversible chromophores allow for the reversible alteration of material properties between two photostationary states and hence these materials can function as molecular switches upon exposure to light of specific wavelengths. Irreversible chromophores, mainly contain photocleavable groups which undergo irreversible changes upon exposure to light, leading to permanent modifications in the polymer structure. Thus, photo-responsive PNCs containing reversible chromophores find applications in fields, such as information storage as a smart polymer (shape memory, self-healing and self-cleaning), artificial muscles, actuators, *etc*., whereas those containing irreversible chromophores can be applied in controlled drug delivery systems, photodegradable materials, photoinduced micropatterning, *etc* [33]. Later, the properties of photo-responsive PNCs can be tuned by controlling factors such as wavelength, intensity, and duration of light exposure; in addition to other factors such as the type, size, and concentration of the chromophoric nanomaterials, as well as the polymer matrix composition. This tunability allows for precise control over the material's response, making it suitable for various applications requiring dynamic modulation of properties.

The photo responsiveness in the PNCs is primarily due to three phenomena; photoisomerization, photothermal effect and photoexcitation [34]. Photoisomerization refers to the reversible transformation of a molecule between two isomeric forms upon absorption of light, which involves the rearrangement of atoms within the molecule's structure, leading to changes in its geometric configuration. This process takes place in polymers containing photoactive groups, such as azobenzene, spiropyran, diarylethene, or fulgides [17]. On the other hand, the photothermal effect involves the conversion of light energy into heat or thermal energy leading to an increase in temperature within the polymer matrix. Zhang *et al*. have designed a photothermal switch composed of Au nanorods (AuNRs) and poly(N-isopropylacrylamide) (PNIPAM) thermosensitive polymers where AuNRs/PNIPAM composite hydrogel showed fast thermal/optical response and structural integrity [35]. Further, the photoexcitation phenomenon involves the absorption of photons by the polymer surface, resulting in the promotion of electrons to higher energy states, due to which it can have various applications in the field of photocatalysis and photoelectricity. In general, nanosized photosensitive semiconductors are incorporated in the PNCs to produce photoelectric materials and photocatalysts. Such types of SRPNCs are used in environmental remediation, water purification, hydrogen production, organic synthesis, and solar energy harvesting. Zhang *et al.* prepared a core-shell structure nanocomposite of γ-Fe_2O_3/PMMA/S-TiO_2 and applied it in the degradation of phenanthrene under visible light irradiation [36].

Thermo-Responsive Polymer Nanocomposites

Thermo-responsive PNCs are a class of materials that exhibit reversible changes in their properties in response to different temperatures. The most common changes include structural and solubility changes resulting in their phase transitions followed by changes in their mechanical properties at specific temperature thresholds. These materials are categorized based on whether their phase transitions between a soluble/swollen state and a collapsed/phase-separated state occur at a lower critical solution temperature (LCST) or at an upper critical solution temperature (UCST). If the phase transition occurs while increasing the temperature, then that phase transition temperature is termed as LCST and if the phase transition occurs on cooling or decreasing the temperature, then that phase transition temperature is termed as UCST. Thus, below LCST or above UCST, the polymer may be soluble or swell in a solvent, while above the LCST or below the UCST, it undergoes phase separation as represented in Fig. (**2**) [37]. The reversible phase separation property of these thermo-responsive polymers makes them an important switchable material with smart properties such as shape memory, self-healing, self-cleaning, anti-fouling, *etc.* with varied applications such as drug delivery, tissue engineering, textile, *etc.*

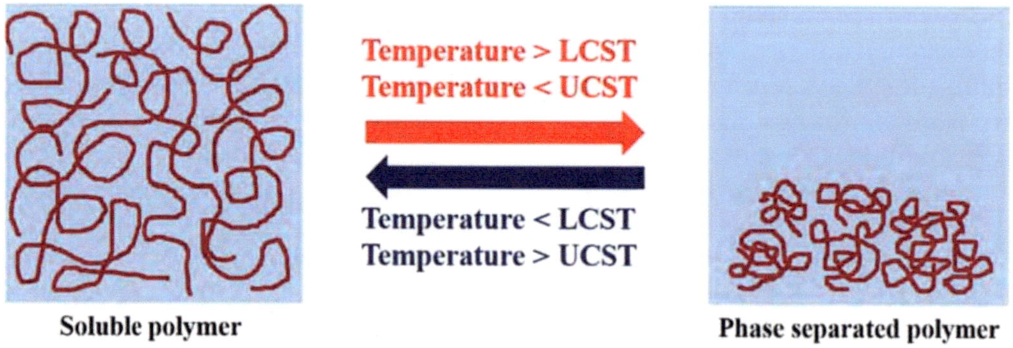

Fig. (2). Representation of phase transition of thermo-responsive PNCs with temperature.

Further, the nanomaterials in PNCs influence the temperature-responsive behaviour of the polymer matrix by exhibiting intrinsic thermo-responsive behaviour, improving heat transfer, increasing surface area, providing mechanical reinforcement, tailoring interfacial properties; and offering functionalization and tunability options. The nanomaterials can exhibit intrinsic thermo-responsive behaviour, which can be synergistically combined with that of the polymer matrix to enhance its thermo-responsiveness [33]. Furthermore, homogeneous dispersion of the nanomaterials within the polymer matrix leading to a controlled morphology can also influence the thermo-responsive behaviour by improved heat transfer and its uniform dissipation throughout the PNC matrix, which is also supported by the increased surface area of nanomaterials offering stronger interactions with the polymer chains. The nanomaterials also provide sufficient mechanical reinforcement, which prevents permanent deformation or phase separation during thermo-responsive transitions [26]. Surface modification of nanomaterials with functional groups also enhances its compatibility with the polymer matrix, leading to more effective transmission of thermal stimuli and improved responsiveness. Later, this functionalization allows for tunability of the thermo-responsive behaviour, enabling the design of PNCs with tailored responsiveness to specific temperature ranges or stimuli.

pH Responsive Polymer Nanocomposites

pH-responsive PNCs represent a versatile class of stimuli-responsive materials that undergo structural and property changes in response to pH changes in the solution. These types of PNC materials are also known as polyelectrolytes since they contain ionizable acidic or basic functional groups within their chemical structure, whose ionization state is sensitive to variations in pH. As these PNCs can undergo ionization by accepting or releasing protons in accordance with their environmental pH, they exhibit behaviours similar to monoacids or monobases. However, the complete ionization of these groups is often hindered by electrostatic effects exerted by adjacent ionized groups within the polymer chain due to which the observable dissociation constant (K_a) of these PNCs differs from that of corresponding monoacids or monobases. This difference in K_a later affects the overall ionization behaviour of the PNCs and their response to pH change [38].

In general, pH-responsive polymers containing basic monomers behave as cationic polymers under acidic conditions, while those containing acidic monomers behave as anionic polymers under basic conditions. Depending on the specific application requirements, it may be necessary to select either cationic or anionic polymers, or a combination of both with the appropriate composition to fabricate the backbone of the photo-responsive polymers [39]. In addition to the introduction of ionizable acidic/basic groups in the polymer backbone, the incorporation of ionizable nanomaterials into the polymer matrix can also induce or enhance the pH responsiveness of the polymers by forming PNCs. The ionizable functional groups in the PNCs undergo ionization when dissolved in a suitable aqueous electrolyte, leading to the formation of charged groups along the polymer chain. These charged groups further experience repulsion in between them, causing the PNC system to expand resulting in the polymer swelling or dissolving as shown in Fig. (**3**). However, if the solvent does not allow for ionization of the PNCs, then the dissolved chain remains in a compact, folded state which later gets precipitated out from the electrolyte solution (Fig. **3**). Further, the addition of more hydrophobic moieties in the PNCs also adds to a more compacted state of the PNCs leading to higher phase separation [38]. This is how pH-responsive PNCs dramatically change conformation in response to minute changes in the ionizing power or pH of the aqueous environment and this behaviour is governed by the interplay between hydrophobic surface energy and electrostatic repulsion between charges. The pH range where reversible phase transitions occur in pH-responsive PNCs can be controlled through two main strategies [40]. The first strategy includes choosing ionizable groups with a pK_a value that matches the desired pH range due to which the ionizable groups become neutral and electrostatic repulsive forces diminish, making the hydrophobic interactions dominant within the polymer network. The other strategy involves the incorporation of hydrophobic moieties into the polymer backbone, which can lead to a more compact conformation of the polymer [41]. By selectively controlling the nature, amount, and distribution of hydrophobic moieties within the polymer backbone, researchers can further modulate the pH responsiveness and phase transition behaviour of pH-responsive polymers. Further, pH-responsive properties of synthetic polymers are generally studied for different applications, however, natural polymers have also been extensively studied for their pH-responsive properties as they offer advantages such as abundance in nature, biodegradability, biocompatibility, and the ability to be modified for specific applications. These attributes make natural polymers particularly attractive for a wide range of biomedical, environmental, and industrial applications [42].

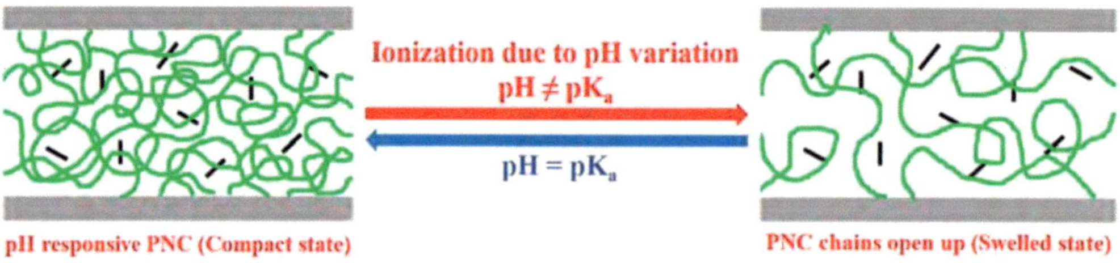

Fig. (3). Representative of conformational change of PNC on pH variation.

Electro Responsive Polymer Nanocomposites

Electro-responsive polymers, also known as electrically responsive or electroactive polymers, are a class of materials that can undergo changes in their shape and volume in response to electrical stimuli. These polymers have the unique ability to transduce electrical energy into mechanical work, making them valuable for various applications in fields such as robotics, actuators, sensors, and biomedical devices. These materials can be categorized into two main types based on the underlying mechanisms of their electro-responsiveness; first as ionic electro-responsive polymers and second as dielectric electro-responsive polymers [43]. The first type exhibits electro-responsiveness due to the mobility of free ions within the polymer matrix in response to an applied electric field. The movement of ions causes changes in the concentration of local ions either within the polymer matrix or in the surrounding solution resulting in an osmotic pressure gradient between the polymer and surrounding solution interface, which later drives the polymer to undergo bending or deformation on the application of mechanical force. These types of polymers include polyelectrolytes, ionomers, and conducting polymers [44]. In contrast, the second type is the type of polymers having a high dielectric constant which exhibits mechanical deformation due to electrostatic (coulombic) forces that develop between two electrodes when the polymer is subjected to an electric field. Dielectric electro-responsive polymers include materials such as ferroelectric polymers, piezoelectric polymers, and electrostrictive polymers [45]. Each type of electro-responsive polymer offers distinct advantages and applications. Ionic polymers are often flexible, lightweight, and capable of large deformations, making them suitable for applications such as soft robotics and artificial muscles. On the other hand, dielectric polymers can exhibit high force generation, making them suitable for applications such as actuators, sensors, and energy harvesting devices. An important aspect to consider is the class of materials that combine electro-responsive polymers with nanomaterials forming PNCs, resulting in enhanced electrical responsiveness compared to traditional polymers. Such electro-responsive PNCs can be fabricated either by using pristine polymers with conductive properties or by using conductive nanomaterials. The high surface area and conductivity of nanomaterials contribute to improved mechanical properties, charge storage, and transfer kinetics, leading to enhanced performance of these materials compared to the pristine polymer matrix. Yang *et al.*, studied a nanocomposite hydrogel of reduced graphene oxide/poly (2-acrylamido-2 methylpropanesulfonic acid-*co*-acrylamide) that possesses excellent electro-responsive and mechanical properties [46]. Another interesting nanocomposite material with application in medical fields was studied by Gangrade *et al.*, where they reported an electro-responsive nanocomposite silk-based drug delivery system containing folic acid-modified single-walled carbon nanotube loaded with the chemotherapeutic drug doxorubicin (DOX) [47].

Magneto-Responsive Polymer Nanocomposites

Magneto-responsive PNCs are another class of materials that exhibit changes in their properties in response to an external magnetic field. These are created by incorporating magnetic nanoparticles or organic ferromagnetic materials into the polymer matrix allowing them to be manipulated or controlled using magnetic fields. Magnetic fields as stimuli offer several advantages over traditional stimuli like temperature and pH, and the most important advantage is that they can penetrate directly allowing for non-contact manipulation of magneto-responsive PNCs. In addition, the magnetic field is non-destructive, exhibits rapid response, and can be easily controlled by adjusting the intensity, direction, and duration of the magnetic field to achieve desired changes in the nanocomposite's behaviour [37]. When PNCs are subjected to a magnetic field, then in response, they may undergo shape deformation, mechanical actuation, magnetic alignment, temperature variation within the PNC matrix, and modulation of properties like mechanical strength, stiffness, and porosity. In the presence of a uniform magnetic field, the magnetophoretic forces acting on the magnetic nanoparticles are transmitted to the polymer matrix in PNC while in the presence of a non-uniform magnetic field, the magnetophoretic force causes magnetic nanoparticles to move towards the regions of stronger magnetic field intensity causing overall distortion of the shape of magneto responsive PNCs [2, 48]. Further, these PNCs also experience a magnetocaloric effect on the application of alternating magnetic field (AMF) where the magnetic nanoparticles generate heat *via* two mechanisms; Neel relaxation and Brownian relaxation. When subjected to an AMF, the magnetic nanoparticles undergo rapid magnetic realignment, leading to frictional losses and the conversion of magnetic energy into heat, which is then transferred to the surrounding polymer matrix. As the polymer matrix absorbs heat, its temperature increases, causing it to swell and release the entrapped triggered molecules [49]. This temperature-induced swelling behaviour makes magnetically-responsive PNCs useful for applications in drug delivery where the swelling can lead to the release of encapsulated drugs in response to magnetic stimuli. Moreover, research has shown that the maximum temperature achieved by the magneto-responsive PNCs increases with the concentration of magnetic nanoparticles and the strength of the externally applied magnetic field. Thus, by adjusting these parameters, the heat generation and temperature response of magnetically responsive PNCs can be tailored for specific applications [50]. Satarkar and Hilt synthesized magnetic nanocomposites by the incorporation of superparamagnetic Fe_3O_4 particles in poly (*N* isopropylacrylamide) hydrogels in order to illustrate remotely controlled drug delivery under a high-frequency alternating magnetic field [51]. Zhao *et al.*, proposed a biocompatible shape memory polymer nanocomposite based on Fe_3O_4 and polylactic acid (PLA) for 4D printing of a tracheal scaffold [52].

APPLICATIONS OF SRPNCS

The integration of versatile polymeric materials with assorted nanostructures produces SRPNCs with multitudinous applications. The ability of these materials to respond differently in different chemical or physical environments triggers their end-use in real-world scenarios as sensors, actuators, electromagnetic shields, water purifiers, drug carriers, *etc*. The following sections demonstrate such reported scenarios where these intelligent SRPNCs were utilized as omni-functional advanced materials.

Sensing and Detection

Sensors are devices or instruments that are designed to recognize any change or abnormality in the environment and to convert these changes into accessible signals. These changes might originate from the deflection in temperature, pressure, light, motion, sound, proximity, concentration, *etc* [53]. Chemical sensors primarily consist of two components; a molecular device that acts as a receptor to identify the analyte molecules and a transducer for transmitting un-amplified signals brought during the interaction of the chemicals into signals [54]. High selectivity, sensitivity, stability, cost-effectiveness, and super detection capability are the prime factors that are considered for developing quality sensors applicable in industry, environmental monitoring, space exploration, biomedicine, pharmaceuticals, *etc* [53, 54]. As stated earlier, because of their inherent features, SRPNCs are the ideal options for designing first-generation sensors. By selecting a suitable polymer matrix, one can achieve the anticipated behaviour while the inclusion of nanostructures further leads to the quantization of energy levels and substantial alteration of the physical and chemical characteristics that synergistically enhances the sensing ability [54].

Metal Ion Sensors

Detection of heavy metals in contaminated water has been practiced extensively using smart PNCs. For instance, the recent study performed by Arjmand *et al.* demonstrated the synthesis of a light-responsive PNC using spiropyran decorated over TiO_2 [55]. The grafting of a novel stimuli-responsive polymer poly(spiropyran ethylacrylate-co-glycidylmethacrylate) over TiO_2 resulted in a decrease of the nanoparticle diameter from 37.2 $m^2.g^{-1}$ to 33.5 $m^2.g^{-1}$. The grafting density was reckoned to be 0.135 chains per nm^2 for the polymer. Interestingly, the photochromatic moiety present over the surface can transform into a merocyanine isomer on exposure to ultra-violet light and can selectively complex with Zn^{2+} ions [55]. A good fatigue resistance of the PNC against Zn^{2+} was recorded making it a portable, high-detecting, and low-cost sensor. In another study, polyvinyl alcohol (PVA) doped sulfur quantum dots (SQDs) are used for the detection of Fe^{3+} and temperature in cells [56]. The hydroxyl-rich PVA polymer has a strong affinity to bind to the surface of SQDs offering them excellent stability. Simultaneously, they can capture Fe^{3+} with a detection limit of 92×10^{-9} M and the fluorescent emission by the SQDs changes upon changing the concentration of the Fe^{3+} ions within 1 min. Similarly, the SQDs appeared to have a temperature-dependent fluorescence with a gradually dropping fluorescent intensity with rising temperature. This feature enabled the use of SQDs as nanothermometers for detecting cell temperatures, at 20 °C and 42 °C. Furthermore, the *in vivo* cell viability study and successful detection of Fe^{3+}were carried out using the SQD strained HeLa cells exposing the cell-loving nature as well as the practical utility of the SQDs for living cell imaging [56]. A porous polyurethane nanocomposite scaffold was capable of detecting chromium ions by using fluorometric interactions [57]. The fluorescence capacity of the base polymer was suitably coupled with the luminescent carbonaceous nanostructures that resulted in the luminescent porous gel. Interestingly, the diffusion of Fe(II), Pb(II), and Cr(VI) ions in 0.1 ppm concentration quenched the luminescence of the gel within 12 h with visibly detected colour change [57].

Gas Sensors

With the increasing environmental depletion due to accelerating lethal emissions into the surroundings, gas sensors are becoming essential for monitoring and controlling toxic gas emissions from various emitters. In this context, intelligent PNCs are exclusively designed and employed as gas sensors because of their high selectivity, optimum sensitivity, large surface area, and good electrical conductivity [58]. Such an example was presented by Sing *et al.* using a polymer-nanostructure-based gas sensor for the detection of highly toxic gas hydrogen cyanide (HCN) [59]. Firstly, the group performed the synthesis of polyaniline (PANI) using oxidative polymerization of aniline in the presence of ammonium persulphate. Then it was integrated with montmorillonite (MMT) clay and reduced graphene oxide (rGO) to prepare PANI/MMT and PANI/MMT-rGO nanocomposite films. The sensors (both PANI/MMT and PANI/MMT-rGO films) were then exposed to 2 ppm concentrations of different gases such as acetone, ammonia, benzene, HCN, and xylene for selectivity where only HCN showed a maximum sensing response of 3.5% and 4.56%, respectively. Further, the sensor response increased up to 9.25% and 11.75%for PANI/MMT and PANI/MMT-rGO, respectively. The study further revealed that the response time of the sensors was quite fast (29.5s for PANI/MMT-rGO and 34.5s for PANI/MMT) and at the same time the films recovered by desorption of HCN within 21s (PANI/MMT) and 25s (PANI/MMT-rGO) [59]. Similarly, another study demonstrated the sensing of ammonia gas by electrochemical polymerization of pyrrole [60]. Polypyrrole films were electrochemically deposited onto indium tin oxide substrates, incorporating metal oxide nanoparticles of MgO and WO_3. The sensing performance of these films towards NH_3 was examined at various operating temperatures and nanoparticle weight percentages. A polyurethane acrylate-multiwalled carbon nanotube nanocomposite was studied as a temperature sensor and H_2, NH_3, and NO_2 adsorption was carried out using Monte Carlo simulation. However, the results obtained from the simulation study supported quality sensing of only NO_2 gas among all [61].

Biomolecule Sensors

For sensing biomolecules like glucose, protein, deoxyribonucleic acid (DNA), antigen, *etc.* smart PNCs with myriad sensibility features can be developed by strategically designing polymer compositions and employing innovative nanostructure-functionalization and synthesis techniques [62]. Ahmed *et al.* demonstrated the same by preparing a set of optical fiber sensors for detecting glucose [63]. Initially, a glucose-sensing hydrogel was fabricated by polymerizing phenyl boronic acid over the tip of a commercial optical fiber. Gold nanoparticles (AuNPs) were separately prepared and deposited over the tip by simply dipping the fiber in the AuNP solution. The variation in surface plasmon resonance (SPR) resulting from the hydrogel's swelling was measured and linked to glucose levels. The sensor's performance was evaluated in both transmission and reflection modes across glucose concentrations typical for healthy individuals (3.9–6.9 mm) and diabetic patients (7.2–10 mm). Additionally, a green laser was utilized to confirm the results obtained through transmission and reflection. The developed sensors exhibited exceptional sensitivity, rapid response (30 s), and a detection limit of 1 mm. Ultimately, a smartphone was utilized to capture alterations in the transmitted light corresponding to varying glucose concentrations, demonstrating the practical application of the developed sensor [63]. Another novel electrochemical sensing platform was prepared from terpolymer/graphene nanosheets for the detection of single-stranded DNA [64]. The polymerization of pyrrole, chlorobenzaldehyde, and heptaldehyde (PPyCB&H) in the presence of graphene nanosheets (GNS) resulted in a PPyCB&H/GNS nanocomposite that was deposited on indium tin oxide substrates for electrochemical detection of single-stranded genomic DNA of *Escherichia coli* (*E. coli*). A more clear understanding of the developments in this field can be obtained from a recent review article, where the authors have showcased the current state-of-the-art examples of SRPNCs-based systems designed for biosensing and bioimaging applications [62]. The study incorporated fresh technological advances in this particular sensing field of biomolecules with a wide range of examples.

From the above discussion, the role of PNCs in designing ideal and quality sensing devices can be realized. However, a comprehensive understanding of PNC properties, including the interaction between the polymer and the nanomaterial components, the influence of PNC size and shape, surface area, sensing mechanism, porosity, and other factors impacting sensing capability, is crucial for controlled synthesis and finest efficiency. Further research is thus essential to develop PNCs as sensing materials that offer improved performance at a reduced cost compared to the existing options [58].

Wastewater Remediation

Water pollution or contamination of water bodies by harmful substances or pollutants originating from various sources, including industrial discharge, agricultural runoff, urban wastewater, and improper disposal of waste materials, *etc*. has become an alarming threat for the ecosystem. Addressing the issues of water pollution requires multi-faceted approaches, among which the improvement of wastewater treatment processes using advanced technologies and materials is an area seeking considerable attention nowadays. Smart PNCs have gathered great traction from researchers as advanced tools for the removal of toxic chemicals including dyes, heavy metal ions, and microorganisms from contaminated water [58]. These intelligent multi-responsive materials demonstrate enhanced properties, including increased resistance to fouling, improved thermal stability, enhanced membrane permeability, superior mechanical characteristics, elevated photocatalytic activity, and enhanced adsorption [58]. The degree of enhancement in these properties varies based on the nanoparticle type, its shape and size, its interaction with the polymer matrix, and its concentration. Additionally, the modification method employed for preparing the PNCs influences their reusability, remediation capacity, and selectivity. Such instances have been exquisitely explored in the recent article by Alipoori *et al.*, where different water contaminants and their removal technologies were reviewed by utilizing polymer-based nanodevices [65].

Dye Removal

Dyes are important synthetic organic compounds utilized by different industries such as petroleum, paper, textile industries, *etc*. There are 100,000 variations of dyes available in the global market and over 1,000,000 tons of dyes are produced every year, the majority comprising the textile dyes. So, the industrial water leakages and washouts containing these dyes massively contribute to water pollution and therefore require exclusive remediation techniques. For such scenarios, PNCs suitably designed with sensitive adsorption or binding sites and porous structures are widely available in the literature and have been utilized for the removal or adsorption of dyes from wastewater. A pH and thermos-responsive poly(*N*-vinylcaprolactum-TiO_2-acrylic acid) based PNC was used for the modification of polysulfone (PSF) membranes with different loading percentages [66]. The membranes were then analysed for various features such as hydrophilic, photocatalytic, pH, thermo-responsive behaviour, *etc*. Most importantly, the modified PSF membrane showed photocatalytic adsorption of methylene blue in dark and UV light exposure. The adsorption activity was tested in three different pH conditions: acidic, basic and neutral pH. The best adsorption profile was recorded under UV light when the medium pH was alkaline in nature, with maximum adsorption of 93.47%. Such results were supposed to arise from the opposite charges present in the membranes and the dye under an alkaline environment. The hydrogel prepared by Sharma *et al.* similarly possessed the capability to remove a set of basic dyes; auramine O, malachite green, and crystal violet with excellent efficiency [67]. The hydrogel was fabricated by blending κ-carrageenan and sodium alginate, grafting polyacrylic acid chains, and crosslinking with *N, N'*-methylenebisacrylamide. To add more, Fe_3O_4 nanoparticles were incorporated into the hydrogel during the synthesis phase to prepare the nanocomposite with better adsorption expertise. The pure hydrogel showed > 95% removal efficiency, however, the presence of the nanoparticles in the PNC increased the removal capacity up to > 99%. A similar PNC-containing iron oxide nanoparticle embedded with PANI exhibited superior adsorption efficiency for Eriochrome Black-T dye [68]. The adsorption capacity of the pure nanoparticles was limited to only low pH conditions while the PNC could adsorb the dye to a varied range of pH (2.0 to 5.0) with excellent competence.

Removal of Microorganisms

Water contamination by pathogenic bacteria poses a significant challenge to human health. Conventional disinfection methods often generate disinfection by-products that can be carcinogenic. Conversely, the currently developed nanotechnologies encounter challenges such as low efficiency and rate, secondary pollution, and complex processes. In this scenario, PNCs serve as the most reliable option for removing these pathogenic organisms without producing any toxic offshoots. Deng *et al.* exemplified the same by fabricating a PNC by functionalizing black phosphorous nanosheets with poly(*N-isopropyl acrylamide*) [69]. The irradiation of the PNC with near-infrared radiation for a short duration facilitated the capture of *E. coli* and *Staphylococcus aureus* (*S. aureus*) bacteria. Prolonged exposure again resulted in the disinfection of the cells due to hyperthermia. Another membrane was developed by Mondal *et al.* that was biocompatible, cost-effective, highly robust, efficient, flexible, freestanding, and reusable [70]. The membrane utilized naturally occurring biocompatible kaolinite clay doped into poly (vinylidene fluoride-hexafluoropropylene) for the effective piezo-dynamic destruction of coliform bacteria. Fig. (**4a**-**c**) show the electron microscopic images of the three PNCs, PVK2.5, PVK5 and PVK7.5 respectively where clay nanoparticles can be seen dispersed in the polymer matrix. Fig. (**4d**-**e**) represent the photographs of the antimicrobial assessment study of PVK5 under ultrasonic treatment (US) against *E. coli* and *E. faecalis* microbes. The bactericidal activity of the membrane was initiated within 10 mins of incubation. While numerous adsorbents and photocatalysts have been developed in recent years, the quest for ideal materials for water treatment applications continues. There is an ongoing effort to develop new sustainable materials that offer high performance at low cost. However, to propel photocatalytic and adsorption-based technologies towards practical implementation, more attention should be given to sustainable SRPNCs to obtain expedient devices.

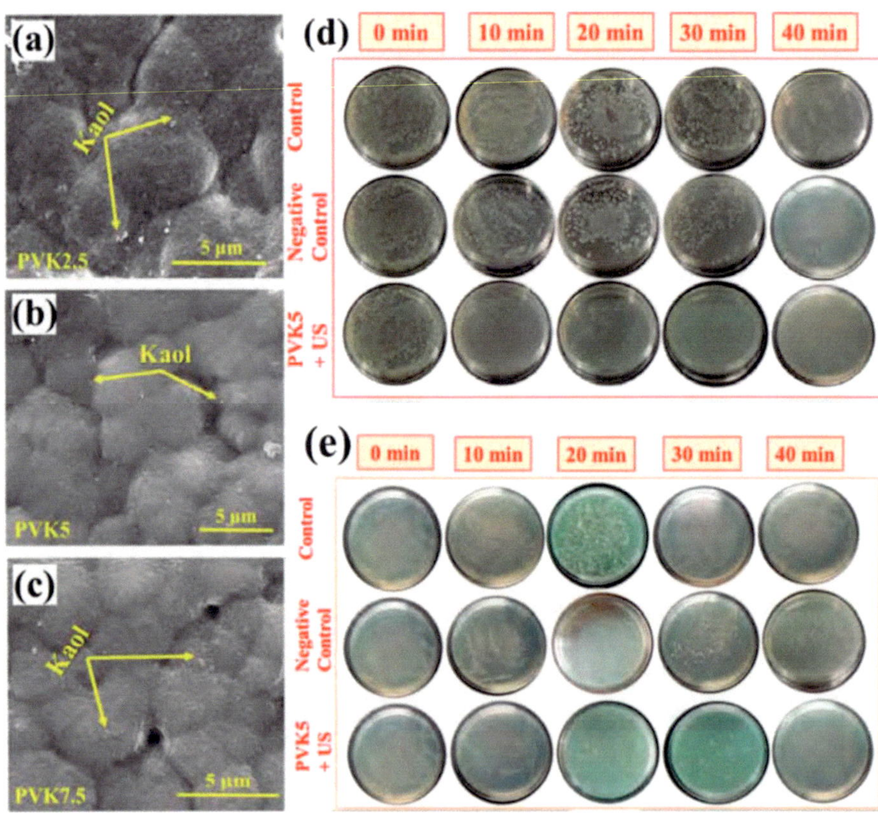

Fig. (4). Field emission scanning electron microscopic images of (**a**) PVK2.5, (**b**) PVK5, and (**c**) PVK7.5; bactericidal test of PVK5 under US treatment for (**d**) *E. coli* and (**e**) *E. faecalis* bacterial colonies (Reproduced form reference [70] with permission, Copyright@2024, ACS).

Metal Ion Removal

The use of SRPNCs for the removal of metal ions during water remediation has gained impetus in recent years. The large surface areas of the nanostructures with variable pore sizes for holding toxic metal ions as well as the flexible polymer matrices with intrinsic functionality capable of coordinating with metal ions make these PNCs suitable candidates for the remediation process. To explain with, a study demonstrated the fabrication of a pH-responsive poly 4-hydroxyphenyl methacrylate single-walled carbon nanotube PNC membrane for the removal Pb^{2+} and Cd^{2+} from wastewater [71]. The adsorbent membrane exhibited a significant adsorption capacity for Pb^{2+} (213.18 mg g^{-1}) and Cd^{2+} (165.37 mg g^{-1}) at an optimal pH of 6.0 at room temperature. The membrane could effectively remove the heavy metal ions in the presence of other interfering ions in up to six repeating cycles [71]. Again, a poly (maleic anhydride)-graft-poly(vinyl alcohol) comb polymer functionalized with magnetic nanoparticles (Fe_3O_4/PMA-g-PVA) was prepared for controlled adsorption of Ag^{2+},Ni^{2+}, Cd^{2+}, Pb^{2+}, and Co^{2+} [72]. The adsorption behaviour confirms more meticulously the Langmuir isotherm model and the pseudo-second-order kinetic model, suggesting monolayer chemisorption. The maximum adsorption capacities were found to be 0.8634, 0.8194, 0.8429, 0.6720, and 0.7682 mmol g^{-1} for Ag^{+}, Ni^{2+}, Cd^{2+}, Pb^{2+}, and Co^{2+}, respectively. Additionally, Fe_3O_4/PMA-g-PVA demonstrates pH responsiveness, facilitating controllable and reversible adsorption/desorption processes. Precisely, adsorption occurs under neutral conditions, while desorption occurs under acidic conditions. This pH-driven mechanism encompasses changes in electrostatic interaction and molecular morphology [72]. Similarly, an alginate-based hydrogel comprising polypyrrole-polyaniline modified GO was employed for the removal of Cr (VI) and Cu (II) from water [73].

Drug Delivery and Tissue Scaffold Systems

Polymer and their derivatives have been known for decades for their ability to competently deliver biomolecules like drugs, proteins, genes, nucleic acids, *etc* [74]. These polymeric nanocarriers have the potential to be tailored into stimuli-responsive systems, operating through triggered release mechanisms. These mechanisms respond to specific stimuli, either endogenous (such as pH, enzymes, temperature, redox values, hypoxia, and glucose levels) or

exogenous (such as light, magnetism, ultrasound, and electrical pulses). This customization enables effective biodistribution and controlled release of drugs or genes at precise target sites. A recent report by Das *et al.* has explained the parameters of SRPNCs and their potential as drug or gene carriers to targeted sites on the basis of novel reported studies [74]. Another review authored by Mertz *et al*. demonstrated the use of different PNCs that can respond to external stimuli and their employment in the delivery of different drug molecules and tissue engineering applications [75]. As discussed earlier, the PNC membrane of poly 4-hydroxyphenyl methacrylate and single-walled carbon nanotube were studied for the delivery of curcumin under controlled pH conditions [69]. For delivering specific drug molecules, the crosslinked structures of hydrogels and their potential swelling attributes are essential. The same was demonstrated by the multi-responsive nanogel/hydrogel nanocomposite, which holds the capacity to deliver levodopa or L-DOPA, a drug used for the treatment of Perkinson's disease [76]. The poly(*N*-isopropylacrylamide) nanogel was incorporated into the hydrogel made from poly (acrylic acid) grafted onto κ-carrageenan polysaccharide and the hydrogel was further crosslinked with magnetic iron oxide nanoparticles. L-DOPA, an amino acid-based drug, shares similar functional groups with the monomers utilized in nanocomposite, facilitating effective interaction with the nanogels and hydrogels. Subsequently, the *in vitro* release of L-DOPA was examined in buffers with pH values of 5.4 and 7.4. The release experiments demonstrated that the release of the L-DOPA drug was accelerated at pH 7.4 and a temperature of 37 °C. Furthermore, the drug release could be sustained for up to 11 days from these soft nanocomposite devices.

Electrically or magnetically responsive PNCs are mandatory for directing magnetic or electrically conducting biomolecules to the specific sites inside the living system. The silica-coated iron oxide/PANI nanocomposite in both nanotubular and granular forms with dual responsive behaviour to magnetic and electric signals demonstrated similar evidence [77]. It was evident that the magnetization values of the nanocomposites showed an increase with higher nanoparticle content. However, electrical conductivity exhibited an inverse relationship. Furthermore, the granular forms displayed significantly higher electrical conductivity (1.1 S/cm) compared to the nanotubular forms, which had a conductivity of 4.8×10^{-2} S/cm. However, the former had a lower saturation magnetization (21 emu/g) compared to the latter (27 emu/g). Another study presented the development of a PNC scaffold composed of arabinoxylan-co-acrylic acid, nano-hydroxyapatite, nano-aluminum oxide, and GO [78]. The silver coating spread over the porous scaffold offered excellent antibacterial properties for *E. coli, S. aureus* and *Pseudomonas aeruginosa* (*P. aeruginosa*) while the biological activity investigated against the mouse preosteoblast cell lines supported biocompatibility of the scaffold in terms of good cell [78]. A magnetic chitosan hydrogel (MCH) was fabricated by incorporating magnetic nanoparticles during the gelation of chitosan that was capable of showing magnetic field-responsive drug release behavior [79]. Fig. (**5a** and **b**) show the transmission electron microscopic (TEM) images of the MCH and magnetic nanoparticles incorporated inside the hydrogel network. Adriamycin (ADM) and rifampicin

(RFP) were used for the study as model drugs. When a low-frequency alternating magnetic field (LAMF) was applied to the MCH, the cumulative release of ADM increased by 67.2% compared to the passive release. From Fig. (**5c** and **d**), a visible rapidity is seen in the cumulative release of the hydrophilic drug ADM than the hydrophobic one. Interestingly, the incorporation of the nanoparticles further improved the mechanical performance as shown in Fig. (**5e** and **f**) [79].

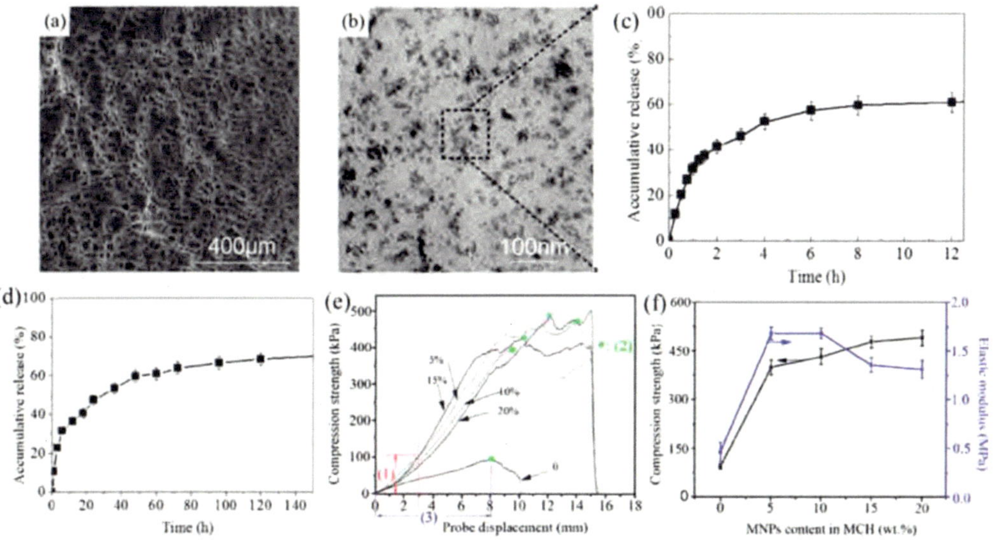

Fig. (5). (**a**) TEM image of MCF and (**b**) high-resolution TEM image of magnetic nanoparticles in MCF; cumulative release of (**c**) ADM and (**d**) RFP loaded MCF; (**e**) compressive strength of MCF with different nanoparticle content; (**f**) compressive and elastic modulus of MCF, (Reproduced with permission from [79], Copyright@2018, ACS).

Although the studies described above exemplify the real-time exploration of PNCs in the field of biomedical and tissue engineering, significant challenges persist in this domain, particularly regarding the attainment of ideal *in vivo* scaffold integration and the realization of remote response mechanisms, both of which remain difficult to achieve [75]. Moreover, there are anticipated future perspectives stemming from these ongoing developments with a never-ending list, some of which have already been initiated.

Smart Packaging for Food Safety

In the food sector, monitoring and sensing technologies play a pivotal role in ensuring the safe preparation and serving of food items. These processes are vital for ensuring that the food we consume is wholesome, uncontaminated, and of the utmost quality. Primarily, it is essential to verify that food is devoid of harmful microorganisms such as bacteria, viruses, and parasites to prevent consumers from falling ill [80]. In this context, the vital role played by SRPNCs cannot be overstated in detecting foreign unwanted micro-species and disinfecting the food. The literature abounds with numerous reports that have already reviewed the recent developments in this field concentrating particularly on food safety-based applications [80 - 82]. Similarly, to combat the significant food wastage observed throughout the food supply chain, research efforts have targeted to deal with the consumer confusion surrounding usability of food items present in the packaging. In response, considerable research has been directed towards developing films with straightforward and easily interpretable indicator systems to detect food spoilage, thereby mitigating the food loss. A recent development in 2021 utilized a glycerol-based dendrimer nanocomposite film as a pH sensor for food packaging [83]. The dyed film was tested for the spoilage of freshly slaughtered pork for a period of 96 hours. It was observed that as the spoilage of the pork started, the pH started to decline and hence the colour of the film started to change indicating a successful release of the dye. The thermomechanical investigation of the films further confirmed their potential to be utilized as food packaging. A multifunctional silver nanoparticle decorated polyaniline and poly(3,4-ethylenedioxythiophene) conducting polymer films were designed for food safety applications [84]. The films can detect the gases liberated by the spoiled food as well as inhibit the formation of biofilms and oxidation of the foodstuffs acting as a protective barrier for them.

When considering food safety, it is always advisable to opt for non-toxic, bioderived polymers, and benign nanostructures to create sustainable PNCs, rather than relying on synthetic polymers and nanostructures. Myriad biopolymers are available in nature and can be utilized in such scenarios to prepare benevolent PNCs. Likewise, a PNC made up of PVA and 15% nanoclay appeared as a tough, flexible and transparent film that showed colorimetric pH indicator and flame-retardant behaviour [85]. The mechanical strength of the films was comparable to the commercial food packaging films made up of low-density polyethylene. The pH-indicative property originated from the anthocyanin, extracted from the red cabbage and a distinct colour change was observed in both acidic and alkaline conditions. Similarly, anthocyanin extracted from barberry and saffron was incorporated in a PNC made up of gelatin and chitosan nanofibers [86]. The film changed its colour upon exposure to different pH levels or ammonia gas and was subsequently used to monitor the freshness of packed fish samples. It was also realized that the antioxidizing and antimicrobial activity of anthocyanin delayed the degradation or spoilage of the fish acting as a protective agent against different microbes.

In spite of these intelligently designed PNCs that are widely available in the literature, the exploration of these materials as food packaging films or containers on the industrial scale is limited. The high production cost and sophisticated methodologies required for the fabrication procedures along with the lack of raw materials for large-scale production can be held responsible for such underrated exploration [58]. Thus, the field urges novel and economical methodologies using abundant materials to widen the practicability of these smart materials.

Smart Materials for Electronic and Automobile Industries

Smart PNCs hold immense promise for energy-related applications, encompassing both energy generation and storage. Commonly employed nanofillers in these applications include metal oxides, nanoclays, carbon nanotubes, and nanosheets. Their exceptional properties, such as high electrical conductivity, environmental stability, thermal conductivity, and mechanical strength, make them highly desirable for modifying the existing energy-related devices and facilitating the development of next-generation smart energy devices usable in automotive as well as other sectors [87]. A recent article on smart PNCs for energy applications described the ongoing trends and developments in the area of using materials such as thermoelectric generators, solar photovoltaic cells, supercapacitors, fuel cells, and hydrogen storage [88].

An exfoliated hexagonal boron nitride-epoxy based PNC was utilized as a thermal interface material for electronic packaging [89]. The excellent thermal conductivity of boron nitride was transferred to the PNC that reduced the thermal resistance in the interface when used as an adhesive between the copper and silicon substrates. In another study by Ponnamma *et al.*, a piezoelectric nanogenerator was fabricated by using the piezoelectric polymer, polyvinylidene fluoride hexafluoropropylene (PVDF-HFP) [90]. The polymer was incorporated with cellulose nanocrystals (CNC) and iron-doped zinc oxide (ZnO) nanoparticles. They were fabricated in the form of double-layered fibers, with CNC/PVDF-HFP on one side and iron-doped ZnO/PVDF-HFP on the other side. From the study, a maximum peak-to-peak output voltage of 12 V and a current density of 1.9 μAcm^{-2}were recorded for the PNC fibers, which was much better than the pure polymer fibers. The nanogenerator was evaluated for its energy harvesting capabilities through exposure to various environments, including ultrasound vibrations and human body movements such as hand tapping and elbow movements. Additionally, it was tested by attaching it to textile fabrics. Results showed that finger tapping generated a peak-to-peak output voltage of 6.5 V, while elbow movements resulted in 5.5 V generation. Overall, the nanogenerator exhibited promising output performance across different movement types, indicating its compatibility with textile materials [90].

The use of PNCs containing various nanostructures in the automobile and aviation industries as lightweight components, corrosion-protecting coatings, electromagnetic interference (EMI) shielding devices, *etc.* can be encountered frequently. Some instances of such developments include poly ethyl ether ketone and ZnO nanocomposite for energy-saving materials [91], polyurethane nanocomposites incorporated with 2-aminothiazole decorated titania-zirconia (TiO_2-ZrO_2) nanoparticles as hydrophobic and corrosion-resistant barrier [92], evenly aligned graphene over epoxy matrix for highly dielectric EMI shielding material [93], *etc.* In a study, the electromagnetic shielding effect of nanocomposites comprising heat-treated carbon nanofibers (CNF) within a linear low-density polyethylene matrix was evaluated [94]. It was observed that heat treatment (HT) of carbon nanofibers at 2500 °C significantly enhanced the intrinsic transport and graphitic crystallinity, resulting in improved shielding efficiency of the nanocomposites. The nanocomposites containing 20 wt.% of heat-treated CNF exhibited a DC electrical conductivity of approximately 1.0 x 10^3 S/m. In the frequency range of 30 MHz to 1.5 GHz, the nanocomposite containing heat-treated CNF demonstrated average shielding effectiveness values of approximately 14 ± 2 dB. Some other examples of PNCs being applied in the fields of electronics, robotics, and automotive industries are shown in Table **1**.

Table 1. Performance of some PNCs in smart electronic and automotive sectors.

Type of Polymer	Nano Structure	Targeted Application	Results	Refs.
Polyurethane	TiO_2/ZrO_2 nanoparticles	Corrosion protection in automobile components	Coating resistance: 1970.17 kÙ.cm^2 Water contact angle: 119°	[92]
Commercial epoxy	Graphene sheets	EMI shielding	Dielectric constant: 14000 EMI shielding efficiency: 38 dB	[93]
Linear low-density polyethylene	Carbon nanofiber	EMI shielding	EMI shielding efficiency: 14 dB	[94]
Leather	Silver nanowire	Joule heating, EMI shielding, piezoresistive sensor	Joule heating temperatures:108 °C EMI shielding efficiency: 55 dB Human motion sensing ability	[95]

(Table 1) cont.....

Type of Polymer	Nano Structure	Targeted Application	Results	Refs.
poly(e-caprolactone)/Thermoplastic polyurethane	Iron oxide@ polydopamine nanoparticles	Light and magnet-responsive soft robotics	Response to light: 5s Response to magnet: 1s Self-healing with light: 120 s Self-healing efficiency: 90.0%	[96]
poly(*N*-isopropyl acrylamide) hydrogel	MXene nanosheets	Soft actuators and sensitive strain sensors	Photothermal responsiveness: 58 °C within 150s Electrical conductivity: 0.35 $S.m^{-1}$	[97]

However, the utilization of PNCs in these cutting-edge sectors is still in the under-explored stage and needs considerable attention from the research fraternity, most importantly to introduce the sustainability concept into the picture.

CONCLUSION

In today's era of technological advancement, where virtually every convenience is embracing smart capabilities, the polymer industry is similarly keeping its pace with these developments to remain competitive. By embracing smart features and leveraging cutting-edge technologies, the industry is poised to meet the challenges and opportunities of the future while delivering high-performance, sustainable solutions to a wide range of applications. SRPNCs epitomize the finest paradigm of progress in this industry. This chapter summarizes a comprehensive exploration of the diverse facets of this realm, including state-of-art synthesis techniques to real-life explorations in various domains of science. The chapter delves into the fabrication techniques of various smart polymers that can respond to different stimuli such as heat, light, pH, and electric and magnetic force. The next part of the chapter describes the employment of these intelligent PNCs in various responsive systems, encompassing sharp sensors, waste-water purifiers, smart food packaging, drug delivery systems, *etc*. The detailed discussion presented in each sector coupled with recent examples clearly pictured the contemporary developments in the field. Nevertheless, the realm of intelligent PNCs has yet to fully embrace the concept of sustainability amidst the ongoing environmental apprehensions. For these, the researchers need to explore nature-based feedstocks for the development of polymer matrices or smart nanomaterials so that the dependency on non-renewable resources can be lowered and more eco-friendly systems can be generated.

REFERENCES

[1] Teyssier, J.; Saenko, S.V.; van der Marel, D.; Milinkovitch, M.C. Photonic crystals cause active colour change in chameleons. *Nat. Commun.*, **2015**, *6*(1), 6368. [http://dx.doi.org/10.1038/ncomms7368] [PMID: 25757068]

[2] Chen, J.K.; Chang, C.J. Fabrications and applications of stimulus-responsive polymer films and patterns on surfaces: A review. *Materials (Basel)*, **2014**, *7*(2), 805-875. [http://dx.doi.org/10.3390/ma7020805] [PMID: 28788489]

[3] Wang, D.; Jin, Y.; Zhu, X.; Yan, D. Synthesis and applications of stimuli-responsive hyperbranched polymers. *Prog. Polym. Sci.*, **2017**, *64*, 114-153. [http://dx.doi.org/10.1016/j.progpolymsci.2016.09.005]

[4] Qureshi, D.; Nayak, S.K.; Maji, S.; Anis, A.; Kim, D.; Pal, K. Environment sensitive hydrogels for drug delivery applications. *Eur. Polym. J.*, **2019**, *120*, 109220. [http://dx.doi.org/10.1016/j.eurpolymj.2019.109220]

[5] Ware, T.; Simon, D.; Rennaker, R.L., II; Voit, W. Smart polymers for neural interfaces. *Polym. Rev. (Phila. Pa.)*, **2013**, *53*(1), 108-129. [http://dx.doi.org/10.1080/15583724.2012.751924]

[6] Hamner, K.L.; Alexander, C.M.; Coopersmith, K.; Reishofer, D.; Provenza, C.; Maye, M.M. Using temperature-sensitive smart polymers to regulate DNA-mediated nanoassembly and encoded nanocarrier drug release. *ACS Nano*, **2013**, *7*(8), 7011-7020. [http://dx.doi.org/10.1021/nn402214e] [PMID: 23899347]

[7] Chen, Y.; Tong, J.; Dong, J.; Luo, J.; Liu, X. A temperature-responsive boronate core cross-linked star (CCS) polymer for fast and highly efficient enrichment of glycoproteins. *Small*, **2019**, *15*(13), 1900099. [http://dx.doi.org/10.1002/smll.201900099] [PMID: 30811830]

[8] Sun, S.; Liang, S.; Xu, W.C.; Xu, G.; Wu, S. Photoresponsive polymers with multi-azobenzene groups. *Polym. Chem.*, **2019**, *10*(32), 4389-4401. [http://dx.doi.org/10.1039/C9PY00793H]

[9] Xu, W.; Hong, Y.; Song, A.; Hao, J. Peptide-assembled hydrogels for pH-controllable drug release. *Colloids Surf. B Biointerfaces*, **2020**, *185*, 110567. [http://dx.doi.org/10.1016/j.colsurfb.2019.110567] [PMID: 31654889]

[10] Fujiwara, N.; Asaka, K.; Nishimura, Y.; Oguro, K.; Torikai, E. Preparation of gold− solid polymer electrolyte composites as electric stimuli-responsive materials. *Chem. Mater.*, **2000**, *12*(6), 1750-1754. [http://dx.doi.org/10.1021/cm9907357]

[11] Zhang, M.; Estournès, C.; Bietsch, W.; Müller, A.H.E. Superparamagnetic hybrid nanocylinders. *Adv. Funct. Mater.*, **2004**, *14*(9), 871-882. [http://dx.doi.org/10.1002/adfm.200400064]

[12] Gao, F.; Pan, X.; Zhu, J.; Zhang, Z.; Zhang, W.; Zhu, X. Facile synthesis of well-defined redox responsive diselenide-labeled polymers *via* organoselenium-mediated CRP and aminolysis. *Polym. Chem.*, **2015**, *6*(8), 1367-1372. [http://dx.doi.org/10.1039/C4PY01481B]

[13] Liu, H.; Rong, L.; Wang, B.; Xie, R.; Sui, X.; Xu, H.; Zhang, L.; Zhong, Y.; Mao, Z. Facile fabrication of redox/pH dual stimuli responsive cellulose hydrogel. *Carbohydr. Polym.*, **2017**, *176*, 299-306. [http://dx.doi.org/10.1016/j.carbpol.2017.08.085] [PMID: 28927612]

[14] Uhlig, K.; Boysen, B.; Lankenau, A.; Jaeger, M.; Wischerhoff, E.; Lutz, J.F.; Laschewsky, A.; Duschl,

C. On the influence of the architecture of poly(ethylene glycol)-based thermoresponsive polymers on cell adhesion. *Biomicrofluidics,* **2012**, *6*(2), 024129.
[http://dx.doi.org/10.1063/1.4729130] [PMID: 23761842]

[15] Xia, Y.; Yin, X.; Burke, N.A.D.; Stöver, H.D.H. Thermal response of narrow disperse poly (N-isopropylacrylamide) prepared by atom transfer radical polymerization. *Macromolecules,* **2005**, *38*(14), 5937-5943.
[http://dx.doi.org/10.1021/ma050261z]

[16] Guragain, S.; Bastakoti, B.P.; Malgras, V.; Nakashima, K.; Yamauchi, Y. Multi-stimuli-responsive polymeric materials. *Chemistry,* **2015**, *21*(38), 13164-13174.
[http://dx.doi.org/10.1002/chem.201501101] [PMID: 26219746]

[17] Manouras, T.; Vamvakaki, M. Field responsive materials: photo-, electro-, magnetic- and ultrasound-sensitive polymers. *Polym. Chem.,* **2017**, *8*(1), 74-96.
[http://dx.doi.org/10.1039/C6PY01455K]

[18] Hsu, L.; Weder, C.; Rowan, S.J. Stimuli-responsive, mechanically-adaptive polymer nanocomposites. *J. Mater. Chem.,* **2011**, *21*(9), 2812-2822.
[http://dx.doi.org/10.1039/C0JM02383C]

[19] Pilate, F.; Toncheva, A.; Dubois, P.; Raquez, J.M. Shape-memory polymers for multiple applications in the materials world. *Eur. Polym. J.,* **2016**, *80*, 268-294.
[http://dx.doi.org/10.1016/j.eurpolymj.2016.05.004]

[20] de Jongh, P.A.J.M.; Mortiboy, A.; Sulley, G.S.; Bennett, M.R.; Anastasaki, A.; Wilson, P.; Haddleton, D.M.; Kempe, K. Dual stimuli-responsive comb polymers from modular N-acylated poly (aminoester)-based macromonomers. *ACS Macro Lett.,* **2016**, *5*(3), 321-325.
[http://dx.doi.org/10.1021/acsmacrolett.5b00904] [PMID: 35614728]

[21] Rivero, R.E.; Molina, M.A.; Rivarola, C.R.; Barbero, C.A. Pressure and microwave sensors/actuators based on smart hydrogel/conductive polymer nanocomposite. *Sens. Actuators B Chem.,* **2014**, *190*, 270-278.
[http://dx.doi.org/10.1016/j.snb.2013.08.054]

[22] Ghahremani Honarvar, M.; Latifi, M. Overview of wearable electronics and smart textiles. *J. Text. Inst.,* **2017**, *108*(4), 631-652.
[http://dx.doi.org/10.1080/00405000.2016.1177870]

[23] Zakerzadeh, E.; Alizadeh, E.; Samadi Kafil, H.; Mohammad Hassanzadeh, A.; Salehi, R.; Mahkam, M. Novel antibacterial polymeric nanocomposite for smart co-delivery of anticancer drugs. *Artif. Cells Nanomed. Biotechnol.,* **2017**, *45*(8), 1509-1520.
[http://dx.doi.org/10.1080/21691401.2016.1260576] [PMID: 27899033]

[24] Bashir, M.; Rajendran, P. A review on electroactive polymers development for aerospace applications. *J. Intell. Mater. Syst. Struct.,* **2018**, *29*(19), 3681-3695.
[http://dx.doi.org/10.1177/1045389X18798951]

[25] Bull, S.J. Multifunctional polymer nanocomposites for industrial applications.*Nanofibers and nanotechnology in textiles*; Woodhead Publishing, **2007**, pp. 256-280.
[http://dx.doi.org/10.1533/9781845693732.2.256]

[26] Liu, T.; Zhou, T.; Yao, Y.; Zhang, F.; Liu, L.; Liu, Y.; Leng, J. Stimulus methods of multi-functional shape memory polymer nanocomposites: A review. *Compos., Part A Appl. Sci. Manuf.,* **2017**, *100*, 20-30.
[http://dx.doi.org/10.1016/j.compositesa.2017.04.022]

[27] Lu, Q.; Jang, H.S.; Han, W.J.; Lee, J.H.; Choi, H.J. Stimuli-responsive graphene oxide-polymer nanocomposites. *Macromol. Res.,* **2019**, *27*(11), 1061-1070.
[http://dx.doi.org/10.1007/s13233-019-7176-3]

[28] Ou, Y.; Yang, F.; Yu, Z.Z. A new conception on the toughness of nylon 6/silica nanocomposite

prepared *via in situ* polymerization. *J. Polym. Sci., B, Polym. Phys.*, **1998**, *36*(5), 789-795. [http://dx.doi.org/10.1002/(SICI)1099-0488(19980415)36:5<789::AID-POLB6>3.0.CO;2-G]

[29] Wang, B.; Peng, D.; Lv, R.; Na, B.; Liu, H.; Yu, Z. Generic melt compounding strategy using reactive graphene towards high performance polyethylene/graphene nanocomposites. *Compos. Sci. Technol.*, **2019**, *177*, 1-9. [http://dx.doi.org/10.1016/j.compscitech.2019.04.013]

[30] Harito, C.; Bavykin, D.V.; Yuliarto, B.; Dipojono, H.K.; Walsh, F.C. Polymer nanocomposites having a high filler content: synthesis, structures, properties, and applications. *Nanoscale,* **2019**, *11*(11), 4653-4682. [http://dx.doi.org/10.1039/C9NR00117D] [PMID: 30840003]

[31] Chow, W.S.; Mohd Ishak, Z.A. Smart polymer nanocomposites: A review. *Express Polym. Lett.*, **2020**, *14*(5), 416-435. [http://dx.doi.org/10.3144/expresspolymlett.2020.35]

[32] Bertrand, O.; Gohy, J.F. Photo-responsive polymers: synthesis and applications. *Polym. Chem.*, **2017**, *8*(1), 52-73. [http://dx.doi.org/10.1039/C6PY01082B]

[33] Shi, Y.; Chen, Z. Function-driven design of stimuli-responsive polymer composites: recent progress and challenges. *J. Mater. Chem. C Mater. Opt. Electron. Devices,* **2018**, *6*(44), 11817-11834. [http://dx.doi.org/10.1039/C8TC02980F]

[34] Lee, W.; Kim, D.; Lee, S.; Park, J.; Oh, S.; Kim, G.; Lim, J.; Kim, J. Stimuli-responsive switchable organic-inorganic nanocomposite materials. *Nano Today,* **2018**, *23*, 97-123. [http://dx.doi.org/10.1016/j.nantod.2018.10.006]

[35] Zhang, C.L.; Cao, F.H.; Wang, J.L.; Yu, Z.L.; Ge, J.; Lu, Y.; Wang, Z.H.; Yu, S.H. Highly stimuli-responsive Au nanorods/poly (N isopropylacrylamide) (PNIPAM) composite hydrogel for smart switch. *ACS Appl. Mater. Interfaces,* **2017**, *9*(29), 24857-24863. [http://dx.doi.org/10.1021/acsami.7b05223] [PMID: 28635250]

[36] Zhang, Y.; Cai, W.; Xie, C.; Wu, H. Visible-light-responsive γ-Fe_2O_3/PMMA/S-TiO_2 core/shell nanocomposite: Preparation, characterization and photocatalytic activity. *Mater. Sci. Semicond. Process.,* **2016**, *48*, 45-51. [http://dx.doi.org/10.1016/j.mssp.2016.02.023]

[37] Yuan, Z.; Ding, J.; Zhang, Y.; Huang, B.; Song, Z.; Meng, X.; Ma, X.; Gong, X.; Huang, Z.; Ma, S.; Xiang, S.; Xu, W. Components, mechanisms and applications of stimuli-responsive polymer gels. *Eur. Polym. J.,* **2022**, *177*, 111473. [http://dx.doi.org/10.1016/j.eurpolymj.2022.111473]

[38] Reyes-Ortega, F. pH-responsive polymers: properties, synthesis and applications. In: *In Smart Polymers and their Applications*; Woodhead Publishing, **2014**; pp. 45-92.

[39] Jiang, T.Y.; Wang, Z.Y.; Tang, L.X.; Mo, F.K.; Chen, C. Polymer micellar aggregates of novel amphiphilic biodegradable graft copolymer composed of poly(aspartic acid) derivatives: Preparation, characterization, and effect of pH on aggregation. *J. Appl. Polym. Sci.,* **2006**, *99*(5), 2702-2709. [http://dx.doi.org/10.1002/app.22818]

[40] Na, K.; Lee, K.; Bae, Y. pH-sensitivity and pH-dependent interior structural change of self-assembled hydrogel nanoparticles of pullulan acetate/oligo-sulfonamide conjugate. *J. Control. Release,* **2004**, *97*(3), 513-525. [http://dx.doi.org/10.1016/S0168-3659(04)00184-1] [PMID: 15212883]

[41] Sukhishvili, S.A. Responsive polymer films and capsules *via* layer-by-layer assembly. *Curr. Opin. Colloid Interface Sci.,* **2005**, *10*(1-2), 37-44. [http://dx.doi.org/10.1016/j.cocis.2005.05.001]

[42] Alvarez-Lorenzo, C.; Blanco-Fernandez, B.; Puga, A.M.; Concheiro, A. Crosslinked ionic

polysaccharides for stimuli-sensitive drug delivery. *Adv. Drug Deliv. Rev.*, **2013**, *65*(9), 1148-1171. [http://dx.doi.org/10.1016/j.addr.2013.04.016] [PMID: 23639519]

[43] Zhang, Q.; Zhang, Y.; Wan, Y.; Carvalho, W.; Hu, L.; Serpe, M.J. Stimuli-responsive polymers for sensing and reacting to environmental conditions. *Prog. Polym. Sci.*, **2021**, *116*, 101386. [http://dx.doi.org/10.1016/j.progpolymsci.2021.101386]

[44] Xu, Z.; Wu, K.; Luo, H.; Wang, Q.; Zhang, T.C.; Chen, X.; Rong, H.; Fang, Q. Electro-responsive SEMI-IPN hydrogel with enhanced responsive property for forward osmosis desalination. *J. Appl. Polym. Sci.*, **2022**, *139*(7), 51650. [http://dx.doi.org/10.1002/app.51650]

[45] Li, Y.; Sun, Y.; Xiao, Y.; Gao, G.; Liu, S.; Zhang, J.; Fu, J. Electric field actuation of tough electroactive hydrogels cross-linked by functional triblock copolymer micelles. *ACS Appl. Mater. Interfaces,* **2016**, *8*(39), 26326-26331. [http://dx.doi.org/10.1021/acsami.6b08841] [PMID: 27617830]

[46] Yang, C.; Liu, Z.; Chen, C.; Shi, K.; Zhang, L.; Ju, X.J.; Wang, W.; Xie, R.; Chu, L.Y. Reduced graphene oxide-containing smart hydrogels with excellent electro-response and mechanical properties for soft actuators. *ACS Appl. Mater. Interfaces,* **2017**, *9*(18), 15758-15767. [http://dx.doi.org/10.1021/acsami.7b01710] [PMID: 28425695]

[47] Gangrade, A.; Gawali, B.; Jadi, P.K.; Naidu, V.G.M.; Mandal, B.B. Photo-electro active nanocomposite silk hydrogel for spatiotemporal controlled release of chemotherapeutics: an *in vivo* approach toward suppressing solid tumor growth. *ACS Appl. Mater. Interfaces,* **2020**, *12*(25), 27905-27916. [http://dx.doi.org/10.1021/acsami.0c02470] [PMID: 32469499]

[48] Nikitin, L.V.; Stepanov, G.V.; Mironova, L.S.; Gorbunov, A.I. Magnetodeformational effect and effect of shape memory in magnetoelastics. *J. Magn. Magn. Mater.*, **2004**, *272-276*, 2072-2073. [http://dx.doi.org/10.1016/j.jmmm.2003.12.838]

[49] Zhang, J.; Huang, Q.; Du, J. Recent advances in magnetic hydrogels. *Polym. Int.*, **2016**, *65*(12), 1365-1372. [http://dx.doi.org/10.1002/pi.5170]

[50] Ang, K.L.; Venkatraman, S.; Ramanujan, R.V. Magnetic PNIPA hydrogels for hyperthermia applications in cancer therapy. *Mater. Sci. Eng. C,* **2007**, *27*(3), 347-351. [http://dx.doi.org/10.1016/j.msec.2006.05.027]

[51] Satarkar, N.; Hilt, J. Magnetic hydrogel nanocomposites for remote controlled pulsatile drug release. *J. Control. Release,* **2008**, *130*(3), 246-251. [http://dx.doi.org/10.1016/j.jconrel.2008.06.008] [PMID: 18606201]

[52] Zhao, W.; Zhang, F.; Leng, J.; Liu, Y. Personalized 4D printing of bioinspired tracheal scaffold concept based on magnetic stimulated shape memory composites. *Compos. Sci. Technol.*, **2019**, *184*, 107866. [http://dx.doi.org/10.1016/j.compscitech.2019.107866]

[53] Norizan, M.N.; Moklis, M.H.; Ngah Demon, S.Z.; Halim, N.A.; Samsuri, A.; Mohamad, I.S.; Knight, V.F.; Abdullah, N. Carbon nanotubes: functionalisation and their application in chemical sensors. *RSC Advances,* **2020**, *10*(71), 43704-43732. [http://dx.doi.org/10.1039/D0RA09438B] [PMID: 35519676]

[54] Sharma, A.; Ahmed, A.; Singh, A.; Oruganti, S.K.; Khosla, A.; Arya, S. Recent advances in tin oxide nanomaterials as electrochemical/chemiresistive sensors. *J. Electrochem. Soc.*, **2021**, *168*(2), 027505. [http://dx.doi.org/10.1149/1945-7111/abdee8]

[55] Arjmand, F.; Mohamadnia, Z. Fabrication of a light-responsive polymer nanocomposite containing spiropyran as a sensor for reversible recognition of metal ions. *Polym. Chem.*, **2022**, *13*(7), 937-945. [http://dx.doi.org/10.1039/D1PY01620B]

[56] Lei, J.; Huang, Z.; Gao, P.; Sun, J.; Zhou, L. Polyvinyl alcohol enhanced fluorescent sulfur quantum dots for highly sensitive detection of Fe^{3+} and temperature in cells. *Part. Part. Syst. Charact.*, **2021**, *38*(4), 2000332.
[http://dx.doi.org/10.1002/ppsc.202000332]

[57] Janghela, S.; Devi, S.; Kambo, N.; Roy, D.; Eswara Prasad, N. Understanding fluorometric interactions in ion-responsive sustainable polymer nanocomposite scaffolds. *Soft Matter,* **2020**, *16*(37), 8667-8676.
[http://dx.doi.org/10.1039/D0SM00965B] [PMID: 32869046]

[58] Darwish, M.S.A.; Mostafa, M.H.; Al-Harbi, L.M. Polymeric nanocomposites for environmental and industrial applications. *Int. J. Mol. Sci.,* **2022**, *23*(3), 1023.
[http://dx.doi.org/10.3390/ijms23031023] [PMID: 35162946]

[59] Singh, A.; Dipak, P.; Iqbal, A.; Samadhiya, A.; Dwivedi, S.K.; Tiwari, D.C.; Tiwari, R.K.; Pandey, K.N. Fast response and recovery polyaniline montmorillonite reduce graphene oxide polymer nanocomposite material for detection of hydrogen cyanide gas. *Sci. Rep.,* **2023**, *13*(1), 8074.
[http://dx.doi.org/10.1038/s41598-023-32151-0] [PMID: 37202389]

[60] Abbas, Y.; Hasan, A. A. Dielectric and gas sensing properties of *in situ* electrochemically polymerized PPy-MgO-WO3 nanocomposite films. *Iraqi Journal of Science,* **2021**, 2915-2933.
[http://dx.doi.org/10.24996/ijs.2021.62.9.8]

[61] Hasanzadeh, I.; Jafari Eskandari, M.; Daneshmand, H. Polyurethane acrylate/multiwall carbon nanotube composites as temperature and gas sensors: Fabrication, characterization, and simulation. *Diamond Related Materials,* **2022**, *130*, 109484.
[http://dx.doi.org/10.1016/j.diamond.2022.109484]

[62] Shu, T.; Shen, Q.; Zhang, X.; Serpe, M.J. Stimuli-responsive polymer/nanomaterial hybrids for sensing applications. *Analyst (Lond.),* **2020**, *145*(17), 5713-5724.
[http://dx.doi.org/10.1039/D0AN00686F] [PMID: 32743626]

[63] Ahmed, I.; El Turk, S.; Al Ghaferi, A.; Samad, Y.A.; Butt, H. Nanocomposite hydrogel-based optical fiber probe for continuous glucose sensing. *Small Sci.,* **2024**, *4*(2), 2300189.
[http://dx.doi.org/10.1002/smsc.202300189]

[64] Pangajam, A.; Chandramohan, A.; Dinakaran, K.; Harichandran, G.; Sureshkumar, R. Preparation and characterization of graphene nanosheets dispersed pyrrole-chorobenzaldehyde-heptaldehyde conjugated terpolymer nanocomposites for DNA detection. *J. Solid State Electrochem.,* **2021**, *25*(10-11), 2611-2623.
[http://dx.doi.org/10.1007/s10008-021-05043-4]

[65] Alipoori, S.; Rouhi, H.; Linn, E.; Stumpfl, H.; Mokarizadeh, H.; Esfahani, M.R.; Koh, A.; Weinman, S.T.; Wujcik, E.K. Polymer-based devices and remediation strategies for emerging contaminants in water. *ACS Appl. Polym. Mater.,* **2021**, *3*(2), 549-577.
[http://dx.doi.org/10.1021/acsapm.0c01171]

[66] Singh, R.; Sinha, M.K.; Purkait, M.K. Stimuli responsive mixed matrix polysulfone ultrafiltration membrane for humic acid and photocatalytic dye removal applications. *Separ. Purif. Tech.,* **2020**, *250*, 117247.
[http://dx.doi.org/10.1016/j.seppur.2020.117247]

[67] Sharma, A.K.; Gupta, A.; Dhiman, A.; Garg, M.; Mishra, R.; Agrawal, G. Fe_3O_4 embedded κ-carrageenan/sodium alginate hydrogels for the removal of basic dyes. *Colloids Surf. A Physicochem. Eng. Asp.,* **2022**, *654*, 130155.
[http://dx.doi.org/10.1016/j.colsurfa.2022.130155]

[68] Deb, A.; Debnath, A.; Saha, B. Sono-assisted enhanced adsorption of eriochrome Black-T dye onto a novel polymeric nanocomposite: kinetic, isotherm, and response surface methodology optimization. *J. Dispers. Sci. Technol.,* **2021**, *42*(11), 1579-1592.
[http://dx.doi.org/10.1080/01932691.2020.1775093]

[69] Deng, S.; Wang, D.; Xiong, Z.; Zhang, S.; Li, D.; Zeng, J.; Zhang, X.; Zhao, Q. Thermo-responsive polymer–black phosphorus nanocomposites for NIR-triggered bacterial capture and elimination. *Environ. Sci. Nano,* **2022**, *9*(4), 1330-1340.
[http://dx.doi.org/10.1039/D1EN00658D]

[70] Mondal, D.; Bag, N.; Roy, J.; Ghosh, S.; Roy, S.; Sarkar, M.; Bardhan, S.; Sutradhar, S.; Das, S. Natural clay-modified piezocatalytic membrane for efficient removal of coliform bacteria from wastewater. *Langmuir,* **2024**, *40*(11), 5785-5798.
[http://dx.doi.org/10.1021/acs.langmuir.3c03560] [PMID: 38446077]

[71] Mamidi, N.; Delgadillo, R.M.V.; Castrejón, J.V. Unconventional and facile production of a stimuli-responsive multifunctional system for simultaneous drug delivery and environmental remediation. *Environ. Sci. Nano,* **2021**, *8*(7), 2081-2097.
[http://dx.doi.org/10.1039/D1EN00354B]

[72] Hwang, L.L., Tseng, H.H. and Chen, J.C., Fabrication of polyphenylsulfone/polyetherimide blend membranes for ultrafiltration applications: The effects of blending ratio on membrane properties and humic acid removal performance. *Journal of membrane science,* **2011**, *384*(1-2), 72-81.

[73] Zhang, W.; Ou, J.; Wang, B.; Wang, H.; He, Q.; Song, J.; Zhang, H.; Tang, M.; Zhou, L.; Gao, Y.; Sun, S. Efficient heavy metal removal from water by alginate-based porous nanocomposite hydrogels: The enhanced removal mechanism and influencing factor insight. *J. Hazard. Mater.,* **2021**, *418*, 126358.
[http://dx.doi.org/10.1016/j.jhazmat.2021.126358] [PMID: 34130162]

[74] Das, S.S.; Bharadwaj, P.; Bilal, M.; Barani, M.; Rahdar, A.; Taboada, P.; Bungau, S.; Kyzas, G.Z. Stimuli-responsive polymeric nanocarriers for drug delivery, imaging, and theragnosis. *Polymers (Basel),* **2020**, *12*(6), 1397.
[http://dx.doi.org/10.3390/polym12061397] [PMID: 32580366]

[75] Mertz, D.; Harlepp, S.; Goetz, J.; Bégin, D.; Schlatter, G.; Bégin-Colin, S.; Hébraud, A. Nanocomposite polymer scaffolds responding under external stimuli for drug delivery and tissue engineering applications. *Adv. Ther. (Weinh.),* **2020**, *3*(2), 1900143.
[http://dx.doi.org/10.1002/adtp.201900143]

[76] Bardajee, G.R.; Khamooshi, N.; Nasri, S.; Vancaeyzeele, C. Multi-stimuli responsive nanogel/hydrogel nanocomposites based on κ-carrageenan for prolonged release of levodopa as model drug. *Int. J. Biol. Macromol.,* **2020**, *153*, 180-189.
[http://dx.doi.org/10.1016/j.ijbiomac.2020.02.329] [PMID: 32135252]

[77] Lalegül-Ülker, Ö.; Elçin, Y.M. Magnetic and electrically conductive silica-coated iron oxide/polyaniline nanocomposites for biomedical applications. *Mater. Sci. Eng. C,* **2021**, *119*, 111600.
[http://dx.doi.org/10.1016/j.msec.2020.111600] [PMID: 33321644]

[78] Khan, M.U.A.; Abd Razak, S.I.; Mehboob, H.; Abdul Kadir, M.R.; Anand, T.J.S.; Inam, F.; Shah, S.A.; Abdel-Haliem, M.E.F.; Amin, R. Synthesis and characterization of silver-coated polymeric scaffolds for bone tissue engineering: antibacterial and *in vitro* evaluation of cytotoxicity and biocompatibility. *ACS Omega,* **2021**, *6*(6), 4335-4346.
[http://dx.doi.org/10.1021/acsomega.0c05596] [PMID: 33623844]

[79] Wang, Y.; Li, B.; Xu, F.; Han, Z.; Wei, D.; Jia, D.; Zhou, Y. Tough magnetic chitosan hydrogel nanocomposites for remotely stimulated drug release. *Biomacromolecules,* **2018**, *19*(8), 3351-3360.
[http://dx.doi.org/10.1021/acs.biomac.8b00636] [PMID: 29995388]

[80] Das, T.K.; Ganguly, S. Revolutionizing food safety with quantum dot–polymer nanocomposites: From monitoring to sensing applications. *Foods,* **2023**, *12*(11), 2195.
[http://dx.doi.org/10.3390/foods12112195] [PMID: 37297441]

[81] Wypij, M.; Trzcińska-Wencel, J.; Golińska, P.; Avila-Quezada, G.D.; Ingle, A.P.; Rai, M. The strategic applications of natural polymer nanocomposites in food packaging and agriculture: Chances, challenges, and consumers' perception. *Front Chem.,* **2023**, *10*, 1106230.

[http://dx.doi.org/10.3389/fchem.2022.1106230] [PMID: 36704616]

[82] Sarker, A.; Mondal, S. C.; Ahmmed, R.; Rana, J.; Ansary, M. W. R.; Bilal, M. Prospects and challenges of polymer nanocomposites for innovative food packaging. *Smart Polymer Nanocomposites,* **2023**, 355-377.
[http://dx.doi.org/10.1016/B978-0-323-91611-0.00021-9]

[83] Pounds, K.; Jairam, S.; Bao, H.; Meng, S.; Zhang, L.; Godinez, S.A.; Savin, D.A.; Pelletier, W.; Correll, M.J.; Tong, Z. Glycerol-based dendrimer nanocomposite film as a tunable pH-Sensor for food packaging. *ACS Appl. Mater. Interfaces,* **2021**, *13*(19), 23268-23281.
[http://dx.doi.org/10.1021/acsami.1c05145] [PMID: 33956422]

[84] El Guerraf, A.; Jadi, S.B.; Ziani, I.; Dalli, M.; Sher, F.; Bazzaoui, M.; Bazzaoui, E.A. Multifunctional smart conducting polymers–silver nanocomposites-modified biocellulose fibers for innovative food packaging applications. *Ind. Eng. Chem. Res.*, **2023**, *62*(11), 4540-4553.
[http://dx.doi.org/10.1021/acs.iecr.2c01327] [PMID: 36975772]

[85] Abu-Thabit, N.; Hakeem, A.S.; Mezghani, K.; Ratemi, E.; Elzagheid, M.; Umar, Y.; Primartomo, A.; Al Batty, S.; Azad, A.K.; Al Anazi, S.; Ahmad, A. Preparation of pH-Indicative and flame-retardant nanocomposite films for smart packaging applications. *Sensors (Basel),* **2020**, *20*(19), 5462.
[http://dx.doi.org/10.3390/s20195462] [PMID: 32977576]

[86] Tavassoli, M.; Alizadeh Sani, M.; Khezerlou, A.; Ehsani, A.; Jahed-Khaniki, G.; McClements, D.J. Smart biopolymer-based nanocomposite materials containing pH-sensing colorimetric indicators for food freshness monitoring. *Molecules,* **2022**, *27*(10), 3168.
[http://dx.doi.org/10.3390/molecules27103168] [PMID: 35630645]

[87] Mansor, M.R.; Akop, M.Z. Polymer nanocomposites smart materials for energy applications.*Polymer nanocomposite-Based smart materials*; Woodhead Publishing, **2020**, pp. 157-176.
[http://dx.doi.org/10.1016/B978-0-08-103013-4.00009-1]

[88] Lin, Z.; Mcnamara, A.; Liu, Y.; Moon, K.; Wong, C.P. Exfoliated hexagonal boron nitride-based polymer nanocomposite with enhanced thermal conductivity for electronic encapsulation. *Compos. Sci. Technol.,* **2014**, *90*, 123-128.
[http://dx.doi.org/10.1016/j.compscitech.2013.10.018]

[89] Li, X.; Xie, Y.; Song, B.; Zhang, H.L.; Chen, H.; Cai, H.; Liu, W.; Tang, Y. A stimuli-responsive smart lanthanide nanocomposite for multidimensional optical recording and encryption. *Angew. Chem. Int. Ed.,* **2017**, *56*(10), 2689-2693.
[http://dx.doi.org/10.1002/anie.201700011] [PMID: 28141905]

[90] Ponnamma, D.; Parangusan, H.; Tanvir, A.; AlMa'adeed, M.A.A. Smart and robust electrospun fabrics of piezoelectric polymer nanocomposite for self-powering electronic textiles. *Mater. Des.,* **2019**, *184*, 108176.
[http://dx.doi.org/10.1016/j.matdes.2019.108176]

[91] Shinde, N. G.; Patel, D. M. Surface modification of polymeric nanocomposites of poly ether ether ketone and zinc oxide for tribological analysis in automobile applications. *Wear, 17*(18), 19.

[92] Xavier, J.R. Electrochemical and dynamic mechanical properties of polyurethane nanocomposite reinforced with functionalized TiO2–ZrO2 nanoparticles in automobile industry. *Appl. Nanosci.,* **2022**, *12*(5), 1763-1778.
[http://dx.doi.org/10.1007/s13204-022-02393-x]

[93] Yousefi, N.; Sun, X.; Lin, X.; Shen, X.; Jia, J.; Zhang, B.; Tang, B.; Chan, M.; Kim, J.K.; Kim, J.K. Highly aligned graphene/polymer nanocomposites with excellent dielectric properties for high-performance electromagnetic interference shielding. *Adv. Mater.,* **2014**, *26*(31), 5480-5487.
[http://dx.doi.org/10.1002/adma.201305293] [PMID: 24715671]

[94] Villacorta, B.S.; Ogale, A.A.; Hubing, T.H. Effect of heat treatment of carbon nanofibers on the electromagnetic shielding effectiveness of linear low density polyethylene nanocomposites. *Polym. Eng. Sci.,* **2013**, *53*(2), 417-423.

[http://dx.doi.org/10.1002/pen.23276]

[95] Ma, Z.; Xiang, X.; Shao, L.; Zhang, Y.; Gu, J. Multifunctional wearable silver nanowire decorated leather nanocomposites for joule heating, electromagnetic interference shielding and piezoresistive sensing. *Angew. Chem. Int. Ed.,* **2022**, *61*(15), e202200705.
[http://dx.doi.org/10.1002/anie.202200705] [PMID: 35122674]

[96] Chen, Y.; Zhao, X.; Li, Y.; Jin, Z.Y.; Yang, Y.; Yang, M.B.; Yin, B. Light- and magnetic-responsive synergy controlled reconfiguration of polymer nanocomposites with shape memory assisted self-healing performance for soft robotics. *J. Mater. Chem. C Mater. Opt. Electron. Devices,* **2021**, *9*(16), 5515-5527.
[http://dx.doi.org/10.1039/D1TC00468A]

[97] Sun, Z.; Song, C.; Zhou, J.; Hao, C.; Liu, W.; Liu, H.; Wang, J.; Huang, M.; He, S.; Yang, M. Rapid photothermal responsive conductive MXene nanocomposite hydrogels for soft manipulators and sensitive strain sensors. *Macromol. Rapid Commun.,* **2021**, *42*(23), 2100499.
[http://dx.doi.org/10.1002/marc.202100499] [PMID: 34480782]

CHAPTER 8

Self-Healing Polymer Nanocomposites and their Applications

Jayanta Madhab Borah[1,*] and **Amlan Puzari**[1]

[1] *Department of Chemistry, Nanda Nath Saikia College, Titabar-785630, Assam, India*

Abstract: Self-healing materials can autonomously repair damage or cracks through dynamic mechanisms. These are designed in such a way that they are associated with a self-repairing mechanism supported by grafted functional groups on the backbone of the polymer. Among the self-healable polymer materials, polymer nanocomposites have drawn much attention in recent years in the growing field of materials science. This chapter is based on the methods of synthesis of self-healing polymer nanocomposites and provides a comprehensive discussion of the applications of self-healing polymer nanocomposites. Owing to their long-term stability and self-healing ability, self-healing polymer nanocomposites are considered one of the best composite materials among the emerging smart materials. Polymer nanocomposites are widely used in electronics, textiles, protective paint and coatings, biomedical, structural materials, *etc*. This chapter is based on the methods of synthesis of self-healing polymer nanocomposites and provides a comprehensive discussion of the applications of self-healing polymer nanocomposites.

Keywords: Biomedical, Construction materials, Electronics, Polymer nanocomposites, Self-healing polymer, Textile.

INTRODUCTION

Among the smart polymeric nanomaterials, self-healing polymer nanocomposites have garnered considerable attention in both research and day-to-day applications [1, 2]. A self-healing material composed of a polymer matrix reinforced with nanoparticles, designed to possess self-healing properties at the nanoscale level. Inspired by nature's resilience and biological systems, self-healing material can autonomously repair damage or cracks through dynamic mechanisms [3].

Self-healing polymer nanocomposites have profound applications in wide-ranging fields such as electronics, protective coatings, biomedical, membranes, structural

* **Corresponding author Jayanta Madhab Borah:** Department of Chemistry, Nanda Nath Saikia College, Titabar-785630, Assam, India; E-mail: jayantamb@gmail.com

Partha Pratim Saikia & Bhaskar Jyoti Saikia (Eds.)

materials, *etc* [4 - 6]. Self-healing polymers are designed in such a way that these are associated with a self-repairing mechanism supported by grafted functional groups on the backbone of the polymer. Such a functionalized polymer forms reversible bonds with different stimuli and possesses a significant advantage in the sense of physical or chemical response toward self-healing [7]. Thus, self-healable polymer nanocomposites have received much attention in recent years in the growing field of materials science [8].

The important characteristics of chemically functionalized polymer nanocomposites lie in their capacity to generate responsive interactions to various stimuli, including light, current, heat, light, *etc.* facilitating self-healing processes [9]. Moreover, self-healing polymer nanocomposites are cost-effective promising alternatives to prepare long-lasting polymer materials and great stands to triumph over environmental issues caused by plastics [10]. Therefore, these materials have considered more durable and eco-friendly materials for R & D and industries. Compared to other conventional materials, self-healing nanomaterials have an extreme ability to self-heal or self-repair when it is subjected to stress-causing strain. The research on self-healing material focused on the discipline of polymer and polymer nanocomposites, which are used in industrial applications and day-to-day life [11, 12].

On the basis of the synthesis and chemistry of self-healing polymer materials, they are classified as (i) autonomic and (ii) non-autonomic self-healing polymers [13, 14]. Autonomic self-healing polymer materials possess the ability to attach chemicals automatically, releaserepair, or heal the response to damage or rupture without any external intervention. The non-autonomic self-healing polymer materials on the other hand, repair or heal the response to damage or rupture in the presence of external stimuli. Resin-based polymers, incorporated as a healing agent in composite materials used in aerospace are non-autonomic materials. Moreover, self-healing polymer materials are also categorized as extrinsic and intrinsic self-healing materials based on their chemistry and applications. Intrinsic polymer materials can repair the damage without external intervention and extrinsic materials can be triggered by the incorporation of external additives to repair the damage or rupture [15].

Self-healing polymer nanocomposites have superior properties like strength, low cost, thermal stability, stiffness, lightweight, *etc.*, which make them potential components in a number of emerging applications [16]. The successful application of self-healing concepts in the manufacture of polymer nanocomposite materials shows a new direction towards sustainability and development in a wide range of fields such as electronics, protective coatings, biomedical, membranes, long-lasting structural materials, *etc* [17].

This chapter discusses the methods of synthesis of self-healing polymer nanocomposites and provides a comprehensive discussion of the applications of self-healing polymer nanocomposites.

DESIGNING OF SELF-HEALING POLYMER NANOCOMPOSITES

A polymeric material shows self-healing properties which are associated with the ability to sense the force causing its damage, and begin the healing process autonomously (without further external stimulus) at the damaged site. A self-healing polymer heals the damage physically or by a combination of physical and chemical processes.

The key steps of all physical processes to design self-healing polymers are the inter-diffusion and entanglement of polymeric chains. Both properties are based on intermolecular forces, which are closely associated with the chemical characteristics of the polymer and the length of the polymeric molecules. Molecular inter-diffusion is induced by Brownian or chain segment motion, entropy-driven movement of particles or molecules, *etc.* in the presence or absence of an external stimulus. Several self-healing mechanisms like welding, patching, swelling, or *via* incorporation of nanoparticles can be assigned to the category proceeding through molecular inter-diffusion.

Among these, the use of nanoparticles to generate self-healing polymeric materials represents an interesting approach to healing different mechanical damages, *viz*. crack (internal damage) or scratch (external damage). The working mechanism of the self-healing polymer nanocomposite materials does not involve breaking and reformation of the polymer backbones like the other physical processes. Instead, it is based on the movement of highly mobile nanoparticles to the damaged vicinity that causes restoring the mechanical strength and thereby healing the damage [18]. In fact, nanoparticles act as dispersants to fill up the damaged part of a polymer causing healing of the material [19]. A small fraction of the dispersed nanoparticles absorb electromagnetic radiation and convert the energy into heat. It results in the formation of new bonds at the damaged site executing self-healing of the material [20 - 23].

On the other hand, the chemical process of designing self-healing polymeric materials is based on covalent or supramolecular type network formation. In covalent network formation, chemical bonds are generated between functional groups resulting in the formation of a permanent or reversible network. On the contrary, supramolecular networks are generally reversible associations of polymers connected through supramolecular interactions [18].

APPLICATIONS OF SELF-HEALING POLYMER NANOCOMPOSITES

The incorporation of nanoparticles in self-healing polymers significantly changes the physical and mechanical properties of the composite material. The combination of both at the nanoscale level results in highly homogeneous novel materials with excellent properties. The real-world applications of self-healing polymer nanocomposites can reduce the maintenance cost, and give long-term durability and safety to the fabricated polymer materials [18]. The self-healing polymer nanocomposites have excellent properties like thermal and electrical conductivity, colour, adhesion, hydrophobicity, reflectivity, *etc.* that play an important role in their application in different fields [19, 20]. Some of the important applications of self-healing polymer nanocomposites are discussed below-

Construction Materials

Self-healing polymer nanocomposites gained attention in construction materials for their ability to autonomously repair damage and thereby increase the service life of structures. The polymer matrix that is reinforced with nanoparticles can enhance the mechanical properties, which facilitate the healing process. Literature reveals that polyvinyl alcohol fibers, super-absorbent polymer, steel, and biochar containing immobilized bacteria are used to detect damages [24].

Immobilized bacteria containing biochar mixed with super-absorbent polymers and fiber have shown significant healing capacity for wider crack surfaces of calcite formed by microbial action. It gives better results formicro-level internal cracks rather than autogenous mechanisms created by the material. In other studies, this has been observed that sulphur-containing super-absorbent polymers combined with Portland cement and calcium sulfoaluminate-based cement have good healing capability, which can quickly heal the damage, sometimes within 30 mins [25].

The thermal shock-resistant cement was reported by mixing poly(ethylene-co-acrylic acid), bisphenol, zinc salt powder, a diglycidyl ether (BPA), and two base cement H. The self-healing ability was retained for 30 days at 300° C for polymer-cement composite materials. Additionally, these advanced cement composite materials show remarkable compressive strength, ductility, mechanical stability, and self-healing properties; for which these can be employed for the extraction of fossil energy [26]. In this way, concrete materials can repair the cracks caused due to shrinkage, weathering, and mechanical loading, thereby increasing the structural integrity, and longevity of buildings and infrastructure.

Textile Industries

Nanocomposites-based self-healing polymeric materials find important textile applications such as protective clothing, workwear, fashionable clothing, *etc*. The self-healing polymeric materials are highly useful for designing protective clothing used for medicalpurposes, the safety of police officers, fire-fighters, fishermen, and factory workers who may face hazards from the environment. The protective action of polymer-based clothing can be improved by using self-regenerating materials in which solution-based easy methods employing self-healing polymer materials are used . For example, a microcapsule-based polymer matrix can be incorporated in textiles, and supramolecular polymers or rubber solutions act as protective films across the cloth surface [27, 28]. Another important application of self-recovering polymer is in making ballistic protective gear that is applied to worn armor along the lines of Kevlar vests and vehicular protection. The ballistic self-healing polymers of poly(ethylene-methacrylic acid) copolymers and ionomers have the ability to self-recover quickly after the impact of a ballistic projectile [29].

The use of self-healing polymer on different textile commodities increases the effectiveness and quality of textile goods. For example, the application of methyl vinyl silicon with inorganic silsesquioxane in making rubber gloves increases the quality and lifespan of gloves. Methyl vinyl silicon rubber and silsesquioxanes upgrade the textile reinforcement by providing good permeation resistance along with barrier protection [30]. The workers working in an environment exposing them to pesticides and chemicals can face detrimental effects on health . Researchers have designed self-healing polymer nanocomposites-based protective clothing to provideprotection from exposure to such chemicals. Researchers have worked on the development of textile products by coating protein, showing high elasticity with polyelectrolyte multilayer (PEM) films. The protein is extracted from squid ring teeth modulus and healing behaviours [31]. The PEMs have prevented the adhesion of bacteria to the material's surface and the deposition of poly-anions and cations on textiles [32]. As a result, these composites neutralize the chemicals exposed to the environment and thereby prevent the human skin.

Another emerging issue in the textile industry is the generation of polluted material day by day in the environment. It may be due to disposing of garments with tears or holes in the fabric [30]. Different approaches are available in the literature to make sustainable designs to increase the lifecycle of clothing and textiles. To overcome these problems, self-healing fabrics can be used in everyday clothes to produce to minimize waste [33]. So, the overall flexibility of the manufacturing method, the use of healing agents, and a blend of multilayer film on the textile are new developments in the clothing industry.

Paint and Coating Materials

The addition of epoxy resins to paint or other coating materials enhances colors and gives a glossy and clean finish. The development of mechanically tough but easily healable and recyclable epoxy resins is highly desirable and beneficial but it remains significantly difficult. Recent development of epoxy resin-based self-healing polymers with considerable toughness has increased the ease of using self-healing polymers for paint and other coating materials. For instance, TiO_2 nanotubes obtained with an average size of 0.02 μm were impregnated with dodecyl amine (act as a corrosion inhibiting agent) and epoxy monomer where TiO_2 nanotubes were spread into the epoxy matrix that could be applied on steel. It was observed that the damaged area of the polymer was practically healed within 96 hrs. Moreover, the corrosion rate of the steel coated with epoxy and healing additives is also observed to be reduced even after immersing the material in saline water for 5 days [34].

Electrical and Electronics

It has been found in the literature that an electronic skin (e-skin) made of nanostructured polybutadiene-based poly(urea-urethane) (PBPUU) is a highly water-non-responsive self-healing elastomer nanocomposite. Its conductive fillers show excellent performance in a range of temperatures, pressure, and pH capable of sensing in both open-air and underwater conditions. Similarly, silver nanowire-based electrodes were successfully used in making multilayered e-skin to extend the heater's connection to the edge. It is reported in the literature that damage-detecting neuron-like composite structure could be shaped using a nanostructured arrangement of conductive carbon black dispersed in the PBPUU which was employed for detecting and mapping local damage. In that material, the sensing layer was made of single-walled carbon nanotubes. It was found that the repair process of the material led to the healing of superficial cracks within 30 seconds. However, healing of extensive damage could require a little bit more time, which is nearly 5 min [35, 36]. Thus, the impregnation of nanostructures into a polymer matrix enhances the overall quality and functionality of the polymer nanocomposite material. These polymers can open up new approaches for the development of smart polymeric materials with high-performance and multifunctional behaviour. However, in making nanocomposite polymer materials, homogenous dispersion of nanomaterials is highly essential to obtain high-performance and better self-healing characteristics [37]. In addition, other factors including morphology of the material, fraction of nanoparticle content and microphase separation are responsible for the self-healing properties of polymer nanocomposites [38].

Miscellaneous

Researchers have found that many polymers with low molecular weight and linked through hydrogen bonding make them robust and induce self-healing properties [39]. These smart self-healable polymeric materials including polymer nanocomposites could be used in various practically useful applications. For instance, smartly designed self-healable glass of polymeric materials has been used to manufacture cell phone screens and other devices with fine structure [40 - 42]. Recently, a photopolymerization-based self-healing photoelastomer was designed comprising a self-healable elastomeric structure that incorporates thiol and disulfide groups. The self-healable polymer material could be used to generate 3D soft actuators, multiphase composites, electronic appliances, *etc* [43]. The development of polymer membranes for separating mixtures of compounds has been a growing area of interest in recent years. Likewise, polyimide nanofibrous membranes associated with the self-healing property, superhydrophobicity, and excellent stability in harsh environments have been developed for efficient separation of oil/water mixtures and water-in-oil type emulsions [44]. A green methodology was developed to synthesize biobased self-healing non-isocyanate polyurethane-urea adduct using carbon dioxide and furfuryl amine as renewable feedstock. The composite was found to have excellent self-healing characteristics, tensile strength, and elongation properties [45]. Self-healing polymer nanocomposites also have diverse applications in the area of biomedicals that include healing of bone, wound, *etc*., artificial skin and dental making, cardiac valve repairing, tissue engineering, anti-bacterial properties *etc* [46 - 48]. For instance, self-healing polymers comprised of selenium are associated with excellent anti-carcinogenic activity that can be used in drug-releasing systems through photo and chemotherapy [49]. Amorphous calcium sulphate nanoparticles composed of resins are induced with dimethylaminohexadecyl methacrylate is extensively used as microcapsules to recover dental tracks. It is highly efficient in remineralization of the crack as well as in calcium delivery. It also has good anti-film, and anti-bacterial properties [50].

CONCLUSION

From the perspectives of long-term stability and self-healing ability, self-healing polymer nanocomposites are considered one of the best composite materials among the emerging smart materials. These smart polymer materials have a wide range of applications in electronics, textiles, protective paint and coatings, biomedical, structural materials, *etc*. Although, the development of self-healing polymer nanocomposites has been done through advanced chemical research, there should be some new innovative inputs for designing and manufacturing smart polymeric nanocomposites according to the practical and industrial need and applications. Thus, from a future perspective, it can be concluded that the self-healing polymer nanocomposites have the advantage of being produced as bulk materials for different practical and emerging fields of the next generation near future that may bring a bright change to numerous areas of research and applications.

REFERENCES

[1] Kessler, M.R.; Sottos, N.R.; White, S.R. Self-healing structural composite materials. *Compos., Part A Appl. Sci. Manuf.,* **2003**, *34*(8), 743-753. [http://dx.doi.org/10.1016/S1359-835X(03)00138-6]

[2] Billiet, S.; Hillewaere, X.K.D.; Teixeira, R.F.A.; Du Prez, F.E. Chemistry of crosslinking processes for self-healing polymers. *Macromol. Rapid Commun.,* **2013**, *34*(4), 290-309. [http://dx.doi.org/10.1002/marc.201200689] [PMID: 23255325]

[3] Chan, Y.L.; Ngan, A.H.; King, N.M. Nanomechanical characterization of mineralized tissues in the oral cavity.*In Emerging Nanotechnologies in Dentistry*; William Andrew Publishing, **2012**, pp. 273-288. [http://dx.doi.org/10.1016/B978-1-4557-7862-1.00016-X]

[4] Yan, L.; Dillard, D.A.; West, R.L.; Rubis, K.J.; Gordon, G.V. Strain rate and temperature dependence of a nanoparticle-filled poly(dimethylsiloxane) undergoing shear deformation. *J. Polym. Sci., B, Polym. Phys.,* **2012**, *50*(13), 929-937. [http://dx.doi.org/10.1002/polb.23088]

[5] Abdullayev, E.; Lvov, Y. Halloysite clay nanotubes as a ceramic "skeleton" for functional biopolymer composites with sustained drug release. *J. Mater. Chem. B Mater. Biol. Med.,* **2013**, *1*(23), 2894-2903. [http://dx.doi.org/10.1039/c3tb20059k] [PMID: 32260855]

[6] Thakur, V.K.; Yan, J.; Lin, M.F.; Zhi, C.; Golberg, D.; Bando, Y.; Sim, R.; Lee, P.S. Novel polymer nanocomposites from bioinspired green aqueous functionalization of BNNTs. *Polym. Chem.,* **2012**, *3*(4), 962-969. [http://dx.doi.org/10.1039/c2py00612j]

[7] Kötteritzsch, J.; Schubert, U.S.; Hager, M.D. Triggered and self-healing systems using nanostructured materials. *Nanotechnol. Rev.,* **2013**, *2*(6), 699-723. [http://dx.doi.org/10.1515/ntrev-2013-0016]

[8] Zheludkevich, M. Self-healing anticorrosion coatings. *Self-healing materials.,* **2009**, 101-39.

[9] Kuhl, N; Bode, S; Hager, MD; Schubert, US Self-healing polymers based on reversible covalent bonds. *Self-healing materials.,* **2016**, 1-58.

[10] Zhang, Z.P.; Rong, M.Z.; Zhang, M.Q. Polymer engineering based on reversible covalent chemistry: A promising innovative pathway towards new materials and new functionalities. *Prog. Polym. Sci.,* **2018**, *80*, 39-93. [http://dx.doi.org/10.1016/j.progpolymsci.2018.03.002]

[11] Guadagno, L.; Vertuccio, L.; Barra, G.; Naddeo, C.; Sorrentino, A.; Lavorgna, M.; Raimondo, M.; Calabrese, E. Eco-friendly polymer nanocomposites designed for self-healing applications. *Polymer (Guildf.),* **2021**, *223*, 123718. [http://dx.doi.org/10.1016/j.polymer.2021.123718]

[12] Sandeep, C.; Harikumar, S.L. Kanupriya,". Hydrogels: a smart drug delivery system. *IJRPC,* **2012**, *2*(3), 603-614.

[13] Orellana, J.; Moreno-Villoslada, I.; Bose, R.K.; Picchioni, F.; Flores, M.E.; Araya-Hermosilla, R. Self-healing polymer nanocomposite materials by Joule effect. *Polymers (Basel),* **2021**, *13*(4), 649.

[http://dx.doi.org/10.3390/polym13040649] [PMID: 33671610]

[14] Wool, R.P. Self-healing materials: a review. *Soft Matter,* **2008**, *4*(3), 400-418. [http://dx.doi.org/10.1039/b711716g] [PMID: 32907199]

[15] Van Tittelboom, K.; Van den Heede, P.; De Belie, N. Self-healing concrete with encapsulated polyurethane.*Eco-Efficient Repair and Rehabilitation of Concrete Infrastructures*; Woodhead Publishing, **2018**, pp. 429-466. [http://dx.doi.org/10.1016/B978-0-08-102181-1.00016-2]

[16] Ling, J.; Rong, M.Z.; Zhang, M.Q. Photo-stimulated self-healing polyurethane containing dihydroxyl coumarin derivatives. *Polymer (Guildf.),* **2012**, *53*(13), 2691-2698. [http://dx.doi.org/10.1016/j.polymer.2012.04.016]

[17] Thakur, V.K.; Vennerberg, D.; Madbouly, S.A.; Kessler, M.R. Bio-inspired green surface functionalization of PMMA for multifunctional capacitors. *RSC Advances,* **2014**, *4*(13), 6677-6684. [http://dx.doi.org/10.1039/c3ra46592f]

[18] *Self-Healing Polymers,*; Ed.Wolfgang H. Binder, Wiley-VCH Verlag GmbH & Co. KGaA, **2013**, pp. 1-53.

[19] Itoh, S.; Kodama, S.; Kobayashi, M.; Hara, S.; Wada, H.; Kuroda, K.; Shimojima, A. Spontaneous crack healing in nanostructured silica-based thin films. *ACS Nano,* **2017**, *11*(10), 10289-10294. [http://dx.doi.org/10.1021/acsnano.7b04981] [PMID: 28957633]

[20] Hu, Z.; Shao, Q.; Huang, Y.; Yu, L.; Zhang, D.; Xu, X.; Lin, J.; Liu, H.; Guo, Z. Light triggered interfacial damage self-healing of poly(p-phenylene benzobisoxazole) fiber composites. *Nanotechnology,* **2018**, *29*(18), 185602. [http://dx.doi.org/10.1088/1361-6528/aab010] [PMID: 29451119]

[21] Qin, W.; Vautard, F.; Askeland, P.; Yu, J.; Drzal, L. Modifying the carbon fiber–epoxy matrix interphase with silicon dioxide nanoparticles. *RSC Advances,* **2015**, *5*(4), 2457-2465. [http://dx.doi.org/10.1039/C4RA11878B]

[22] Hu, Z.; Zhang, D.; Yu, L.; Huang, Y. Light-triggered C_{60} release from a graphene/cyclodextrin nanoplatform for the protection of cytotoxicity induced by nitric oxide. *J. Mater. Chem. B Mater. Biol. Med.,* **2018**, *6*(3), 518-526. [http://dx.doi.org/10.1039/C7TB02624B] [PMID: 32254531]

[23] Huang, L.; Yi, N.; Wu, Y.; Zhang, Y.; Zhang, Q.; Huang, Y.; Ma, Y.; Chen, Y. Multichannel and repeatable self-healing of mechanical enhanced graphene-thermoplastic polyurethane composites. *Adv. Mater.,* **2013**, *25*(15), 2224-2228. [http://dx.doi.org/10.1002/adma.201204768] [PMID: 23417742]

[24] Thakur, V.K.; Kessler, M.R. Synthesis and characterization of AN- *g* -SOY for sustainable polymer composites. *ACS Sustain. Chem.& Eng.,* **2014**, *2*(10), 2454-2460. [http://dx.doi.org/10.1021/sc500473a]

[25] Lin, M.F.; Thakur, V.K.; Tan, E.J.; Lee, P.S. Dopant induced hollow BaTiO3 nanostructures for application in high performance capacitors. *J. Mater. Chem.,* **2011**, *21*(41), 16500-16504. [http://dx.doi.org/10.1039/c1jm12429c]

[26] Sánchez-Romate, X.F., Sans, A., Jiménez-Suárez, A. and Prolongo, S.G., The addition of graphene nanoplatelets into epoxy/polycaprolactone composites for autonomous self-healing activation by Joule's heating effect. *Composites Science and Technology,* **2021**, *213*, 108950.

[27] Kua, H.W.; Gupta, S.; Aday, A.N.; Srubar, W.V., III Biochar-immobilized bacteria and superabsorbent polymers enable self-healing of fiber-reinforced concrete after multiple damage cycles. *Cement Concr. Compos.,* **2019**, *100*, 35-52. [http://dx.doi.org/10.1016/j.cemconcomp.2019.03.017]

[28] Gwon, S.; Ahn, E.; Shin, M. Water permeability and rapid self-healing of sustainable sulfur composites using superabsorbent polymer and binary cement. *Constr. Build. Mater.,* **2020**, *265*,

120306.
[http://dx.doi.org/10.1016/j.conbuildmat.2020.120306]

[29] Koech, P.K.; Fernandez, C.A.; Rod, K.A.; Dai, G.; Huerta, N.J.; Burton, S.; Miller, Q.R.; Resch, C.T. Advanced self-healing polymer-cement composites for geothermal wellbore applications up to 300 C. *PROCEEDINGS World Geothermal Congress,* Reykjavik, Iceland**2020**.

[30] Heyse, P.; De Vilder, I.; Vanneste, M. Smart durable and self-healing textile coatings.*In Active Coatings for Smart Textiles*; Hu, J.L., Ed.; Woodhead Publishing Series in Textiles, Woodhead Publishing, **2016**, pp. 55-80.
[http://dx.doi.org/10.1016/B978-0-08-100263-6.00004-6]

[31] Dannehl, A.; Buhr, A.; Sanchez Leyton, A.; Hellweg, L.; Beer, M.; Sabantina, L. Self-healing materials for potential use in textile and clothing applications. *Communications in Development and Assembling of Textile Products,* **2023**, *4*(1), 27-41.
[http://dx.doi.org/10.25367/cdatp.2023.4.p27-41]

[32] Kalista, S.J.; Pflug, J.R.; Varley, R.J. Effect of ionic content on ballistic self-healing in EMAA copolymers and ionomers. *Polym. Chem.,* **2013**, *4*(18), 4910-4926.
[http://dx.doi.org/10.1039/c3py00095h]

[33] Iacono, S.D.; Martone, A.; Amendola, E. *Diels-Alder Chemistry to Develop Self-Healing Epoxy Resins and Composites Thereof*; Paint and Coatings Industry, **2018**.
[http://dx.doi.org/10.5772/intechopen.81360]

[34] Ubaid, F.; Radwan, A.B.; Naeem, N.; Shakoor, R.A.; Ahmad, Z.; Montemor, M.F.; Kahraman, R.; Abdullah, A.M.; Soliman, A. Multifunctional self-healing polymeric nanocomposite coatings for corrosion inhibition of steel. *Surf. Coat. Tech.,* **2019**, *372*, 121-133.
[http://dx.doi.org/10.1016/j.surfcoat.2019.05.017]

[35] Khatib, M.; Zohar, O.; Saliba, W.; Srebnik, S.; Haick, H. Highly efficient and water-insensitive self-healing elastomer for wet and underwater electronics. *Adv. Funct. Mater.,* **2020**, *30*(22), 1910196.
[http://dx.doi.org/10.1002/adfm.201910196]

[36] Khatib, M.; Zohar, O.; Saliba, W.; Haick, H. A multifunctional electronic skin empowered with damage mapping and autonomic acceleration of self-healing in designated locations. *Adv. Mater.,* **2020**, *32*(17), 2000246.
[http://dx.doi.org/10.1002/adma.202000246] [PMID: 32173928]

[37] Binder, W.H., Ed. *Self-healing polymers: from principles to applications*; John Wiley & Sons, **2013**.
[http://dx.doi.org/10.1002/9783527670185]

[38] Iyer, B.V.S.; Yashin, V.V.; Hamer, M.J.; Kowalewski, T.; Matyjaszewski, K.; Balazs, A.C. Ductility, toughness and strain recovery in self-healing dual cross-linked nanoparticle networks studied by computer simulations. *Prog. Polym. Sci.,* **2015**, *40*, 121-137.
[http://dx.doi.org/10.1016/j.progpolymsci.2014.07.004]

[39] Yanagisawa, Y.; Nan, Y.; Okuro, K.; Aida, T. Mechanically robust, readily repairable polymers *via* tailored noncovalent cross-linking. *Science,* **2018**, *359*(6371), 72-76.
[http://dx.doi.org/10.1126/science.aam7588] [PMID: 29242235]

[40] Rahman, M.W.; Shefa, N.R. Minireview on self-healing polymers: versatility, application, and prospects. *Adv. Polym. Technol.,* **2021**, *2021*, 1-12.
[http://dx.doi.org/10.1155/2021/7848088]

[41] Zhou, Y.; Li, L.; Han, Z.; Li, Q.; He, J.; Wang, Q. Self-healing polymers for electronics and energy devices. *Chem. Rev.,* **2023**, *123*(2), 558-612.
[http://dx.doi.org/10.1021/acs.chemrev.2c00231] [PMID: 36260027]

[42] Atay, HY Self-healing composite materials. *Aerospace Polymeric Materials.,* **2022**, *1*, 137-54.
[http://dx.doi.org/10.1002/9781119905264.ch6]

[43] Yu, K.; Xin, A.; Du, H.; Li, Y.; Wang, Q. Additive manufacturing of self-healing elastomers. *NPG*

Asia Mater., **2019**, *11*(1), 7.
[http://dx.doi.org/10.1038/s41427-019-0109-y]

[44] Ma, W.; Ding, Y.; Li, Y.; Gao, S.; Jiang, Z.; Cui, J.; Huang, C.; Fu, G. Durable, self-healing superhydrophobic nanofibrous membrane with self-cleaning ability for highly-efficient oily wastewater purification. *J. Membr. Sci.,* **2021**, *634*, 119402.
[http://dx.doi.org/10.1016/j.memsci.2021.119402]

[45] Wu, P.; Cheng, H.; Wang, X.; Shi, R.; Zhang, C.; Arai, M.; Zhao, F. A self-healing and recyclable polyurethane-urea Diels–Alder adduct synthesized from carbon dioxide and furfuryl amine. *Green Chem.,* **2021**, *23*(1), 552-560.
[http://dx.doi.org/10.1039/D0GC03695A]

[46] Miao, S.; Castro, N.; Nowicki, M.; Xia, L.; Cui, H.; Zhou, X.; Zhu, W.; Lee, S.; Sarkar, K.; Vozzi, G.; Tabata, Y.; Fisher, J.; Zhang, L.G. 4D printing of polymeric materials for tissue and organ regeneration. *Mater. Today,* **2017**, *20*(10), 577-591.
[http://dx.doi.org/10.1016/j.mattod.2017.06.005] [PMID: 29403328]

[47] Pike, F.S.; Kumar, M.S.A.; Boudrieau, R.J. Reduction and fixation of cranial cervical fracture/luxations using screws and polymethylmethacrylate (PMMA) cement: a distraction technique applied to the base of the skull in thirteen dogs. *Vet. Surg.,* **2012**, *41*(2), 235-247.
[http://dx.doi.org/10.1111/j.1532-950X.2011.00916.x] [PMID: 22188100]

[48] Hou, C.; Huang, T.; Wang, H.; Yu, H.; Zhang, Q.; Li, Y. A strong and stretchable self-healing film with self-activated pressure sensitivity for potential artificial skin applications. *Sci. Rep.,* **2013**, *3*(1), 3138.
[http://dx.doi.org/10.1038/srep03138] [PMID: 24190511]

[49] Xia, J.; Li, T.; Lu, C.; Xu, H. Selenium-containing polymers: perspectives toward diverse applications in both adaptive and biomedical materials. *Macromolecules,* **2018**, *51*(19), 7435-7455.
[http://dx.doi.org/10.1021/acs.macromol.8b01597]

[50] Wu, J.; Weir, M.D.; Zhang, Q.; Zhou, C.; Melo, M.A.S.; Xu, H.H.K. Novel self-healing dental resin with microcapsules of polymerizable triethylene glycol dimethacrylate and N,N-dihydroxyethyl-p-toluidine. *Dent. Mater.,* **2016**, *32*(2), 294-304.
[http://dx.doi.org/10.1016/j.dental.2015.11.014] [PMID: 26743969]

CHAPTER 9

Smart Biopolymers and their Applications

Rajesh Mandal[1] and **Arup Kumar Dutta**[1,*]

[1] *Department of Chemistry, Pandit Deendayal Upadhyaya Adarsha Mahavidyalaya, Behali, Assam 784184, India*

Abstract: Over the last few decades, polymers have gained lots of importance in everyday life due to their unique properties and wide range of applications. In recent years, biopolymers have received attention due to their features like biodegradability, biocompatibility, renewability, and inexpensiveness. They are also used as smart and active materials in medicine, the food industry, *etc*., as they exhibit responsiveness to factors like temperature, pH, humidity, light, electrical and magnetic fields, *etc*. Therefore, this chapter emphasizes an in-depth analysis of various techniques and methodologies employed in the preparation of smart biopolymers, their characterization, applications, and the challenges associated with their development and utilization.

Keywords: Active-biopolymer, Bioengineering, Drug delivery, Edible coating, Functional food, Plastic pollutions, Protein nanoparticles, Renewable materials, Sustainable materials, Smart-biopolymers, Tissue engineering.

INTRODUCTION

Biopolymers, or polymeric biomolecules, are mostly biologically degradable polymers and are produced using living organisms, especially plants. Biopolymers are usually renewable and sustainable materials as they are prepared from living organisms. Their monomeric units are covalently bonded together, which leads to larger molecular structures [1 - 3]. Plants, microbes, and other naturally available materials are major sources of biopolymers. Different renewable materials are used to produce varieties of biopolymers that possess various physical, chemical, and biochemical properties [3 - 5]. In our nature, we have varieties of naturally available biopolymers such as lignin, starch, cellulose, hemicelluloses, *etc*. Biopolymers that are developed from natural resources have been manufactured in large quantities [6]. The biodegradable nature of biopolymers is due to the presence of some specific enzymes as well as microorganisms such as bacteria, algae, and fungi that possess distinct degradable characteristics [3]. The biopo-

* **Corresponding author Arup Kumar Dutta:** Department of Chemistry, Pandit Deendayal Upadhyaya Adarsha Mahavidyalaya, Behali, Assam 784184, India; E-mail: arupdutta35@gmail.com

Partha Pratim Saikia & Bhaskar Jyoti Saikia (Eds.)

lymers have oxygen and nitrogen atoms in their polymeric chain, which make their biodegradability very simple [7, 8]. During degradation, these biopolymers are converted into biomass, water, carbon dioxide, and other natural components. The biopolymers are responsible for holding cells together in tissues, providing signals to the cells [9 - 11]. They are also responsible for providing elasticity and hydration to the skin and lubricating gastrointestinal tracts as well as joints of our body. Due to their additional benefits and functionality, the biopolymer possesses anadvantage over traditional polymers. There are many applications of biopolymers. These biopolymers can be utilized for reducing the simulation of chronic inflammation, and other toxic effects in the human body [12]. Biopolymers, being biodegradable, also have a variety of applications in the field of bioengineering like tissue engineering, drug delivery, and wound dressing. They can also be used in food packaging material, municipal solid waste disposal that reduces plastic pollution, and emission of carbon dioxide, resulting in a sustainable environment [13 - 15]. These biopolymers are often termed smart or adaptable materials, as their structure can be manipulated by using enzymes at different stages of their lifecycle. Moreover, these smart biopolymers are sensitive to small changes in the environment in terms of temperature, pH, solvents, *etc.*, and may undergo fast and reversible changes in their structure and hence also in their physical properties [16 - 19].

Biopolymers: A State of Art

Based on origin, biopolymers are classified into two major categories. These are natural biopolymers, and synthetic biopolymers. These biopolymers are mainly found in living organisms like plants, animals, microbes, *etc.*

Natural Biopolymers

Biopolymers that are extracted from living organisms are known as natural biopolymers; for example, polysaccharides and proteins. Polysaccharides are long-chain carbohydrate molecules, or monosaccharides connected through glycosidic linkage. The major functions of polysaccharides in the human body are energy storage and structural support. They are abundant natural biopolymers found in animals, plants, and other microorganisms. Starch, cellulose, chitosan, chitin, *etc.* belong to naturally occurring important polysaccharides. Similarly, proteins also belong to an important class of biopolymers that perform a number of important functions in the cells of the human body. They play an important role in the tissues and organs of the human body in terms of construction, function, *etc.* They are made of monomeric units known as amino acids that are linked by peptide bonds. Proteins are also known as polypeptides [20 - 22]. Proteins are also hydrophilic in nature as they are made up of polar amino acid monomers. In a protein, several different amino acids may be linked together. Plant proteins are more immunogenic than animal proteins, have low molecular weight, and are mainly found in foods like corn, wheat, soybeans, *etc* [23].

Synthetic Biopolymer

Synthetic biopolymers are biological polymers that are either synthesized in the laboratory/industry or are prepared by modifying natural polymers. Synthetic biopolymers have several advantages over natural polymers in terms of their stability, and flexibility in various applications, and that is why synthetic biopolymers have been receiving lots of attention in the past decades. Moreover, synthetic biopolymers are preferable compared to synthetic polymers, as most of the synthetic biopolymers are biodegradable [23 - 25]. In recent years, synthetic biopolymers have been widely utilized in the medicine industry because of their specific properties like stability, controlled release, easy removal/degradability, and less or no immunogenicity. Synthetic biopolymers can be classified into the following two categories:

i. Biodegradable biopolymers and
ii. Non-biodegradable biopolymers.

Biodegradable polymers disintegrate by themselves due to the natural degradation through the action of microorganisms over some time. Due to their biodegradable nature, biopolymers become a viable alternative for packaging materials. Synthetic biodegradable polymers are more useful due to their mechanical properties. They also play an important role in tissue engineering. On the other hand, non-biodegradable biopolymers are a kind of biopolymer that cannot be broken down naturally by the action of microorganisms and end up as garbage, or a source of pollution [26 - 28]. They are mostly used in non-biomedical applications. The common examples are PA, PMMA, polycarbonate, PU, *etc*.

Classification of Smart Biopolymer

In the previous section, we discussed biopolymers and their classifications. Certain biopolymers tend to respond in the desired way due to small changes in pH, temperature, *etc*. known as smart biopolymers. In recent years, several smart biopolymeric materials have been developed as there is exponential growth in the application of smart biopolymers. Based on functionality and types of stimuli, smart biopolymers are classified into the following categories.

pH-sensitive Smart Polymers

Polymers that can respond to a change in pH are termed pH-sensitive smart polymers. They possess ionizable acidic (carboxylic or sulphonic), or basic (amino salt) functional groups in their structure. The carboxylic groups of the polymer solution accept protons in poly-anionic form at low pH and release protons at high pH to maintain the pH [29 - 31]. Examples of polyanions are PAA [poly(acrylic acid)], PMAA [poly(methacrylic acid)], *etc.* On the other hand, polymers with basic functional groups accept protons during high pH in polybasic form and are ionized at low pH. Examples are PVP [poly(2-vinyl pyridine)], PDEAEMA [poly(2-diethyl aminoethyl methacrylate)], *etc.*

Thermo-sensitive Smart Polymers

These are the smart polymers that change their structural features with a temperature change. These smart polymers exhibit critical solution temperature points above or below which the polymer is completely miscible with solvents, and the points are known as upper critical solution temperature (UCST) or lower critical solution temperature (LCST) respectively. In case of polymers with LCST when the temperature is increased, the polymer is precipitated out of the polymer solution, and therefore they are widely used in drug delivery systems [32 - 34]. Examples of thermo-sensitive polymers with LCST are poly(N,N-diethyl acrylamide), poly(N-vinylalkylamide), PNIPAAm [poly(N-isopropyl acrylamide], *etc.*

Stimuli-Responsive Smart Polymers

These are the polymers that are sensitive to exposure to different types of stimuli like light, electric current, magnetic fields, *etc.* These polymers can undergo phase transition or change in physical properties when exposed to the stimuli. These types of polymers are used in industrial separation techniques, in cancer therapy agents, and sometimes in drug delivery systems [35 - 37].

APPLICATIONS OF SMART BIOPOLYMER

In recent decades, the application of smart biopolymers has increased by manyfold. They have broad and important applications in the fields of pharmaceuticals and food industries (Fig. **1**). Some of their important applications are discussed in the chapter.

Smart Biopolymers for Medical Applications

In the last few decades, biopolymers played a major role in drug delivery systems that increase effectiveness in drug delivery. Synthetic, and natural biopolymers are commonly used for this purpose. The major applications of biopolymers in drug delivery systems are controlled drug administration, suturing, fixing, adhesion, occlusion, isolation, contact inhibition, *etc* [38 - 40]. Due to a very high degree of adaptability, degradation efficiency, and unique mechanical characteristics, biopolymers are considered innovative materials. These biopolymers have become an alternative to commonly used synthetic polymers considering their biodegradable nature and lack of cytotoxicity. Most of the biopolymers are synthesized from natural sources through an enzyme-catalyzed polymerization process, and some of them even develop within cells during the cell metabolism process [41, 42]. A few examples of such types of biopolymers are collagen, gelatin, cellulose, alginate, hyaluronic acid, *etc*. One of the important features of the biodegradable polymer in drug delivery systems is that after the release of the medicine, no surgery is needed to remove it. These biodegradable polymers have been utilized in some major biomedical applications like soft-tissue replacement, coatings for tablets & capsules, breast implants, vascular grafting, blood substitution, cardiac assist devices, pacemakers, *etc*. They also have applications in contact lenses, livers, bladders, kidneys, catheters, heart valves, *etc* [43 - 45]. Nowadays, nanoparticles based on biodegradable biopolymers have been successfully utilized for potential biomedical applications. These nano biopolymers, based on their specifications and also owing to their unique characteristics like biocompatibility,and safety, possess distinct therapeutic efficacy and bioavailability [46].

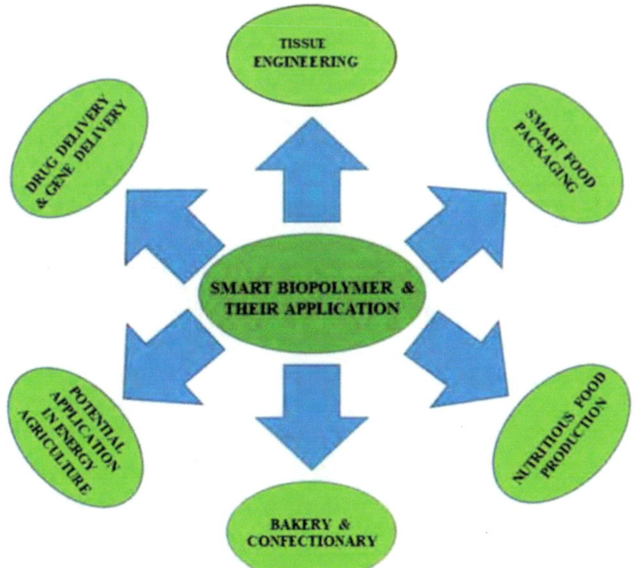

Fig. (1). Application of smart biopolymers in various fields.

Among biopolymers, protein and polysaccharide-based polymers play a vital role in biomedical applications. Due to the molecular size of protein molecules, they can be easily converted into protein nanoparticles. These biodegradable protein nanoparticles have potential use in tissue engineering that increases the healing potential in injured or damaged tissues [53 - 55]. These protein nanomaterials are also utilized in therapeutic medicine/chemical delivery, like drug delivery, gene delivery, *etc*., and for this purpose, collagen protein-based nanoparticles, electro-spun fibers, sponges, and hydrogels are used. Other protein-based biomolecules that are used in biomedical applications are gelatin, keratin, sericin, fibroin, *etc*. These proteins are used in the form of emulsions, nanofibers, films, hydrogels, 3D-printed scaffolds, *etc*., in different biomedical applications [56 - 59]. Like protein biopolymers, polysaccharides are also highly biodegradable, and biocompatible. Chitosan, a linear polysaccharide, plays an important role in cell growth and attachment. Chitosan in its porous structure is widely used in tissue engineering. Chitosan in combination with other materials is broadly used in bone and ligament implantation and also in skin generation [60, 61]. However, chitosan-based nanomaterials are not convincing enough for biomedical applications owing to their poor stability and sensitivity towards temperature and pH, even though they possess characteristics like less toxicity, high permeability, and high moisture retention capacity [62 - 64]. Some of the important biopolymers, along with their medical applications, are presented in Table **1**.

Table 1. Example of biopolymer along with biomedical application.

Biopolymers	Biomedical Application (Refs.)
Collagen	Surface coating for tissue culture plates [47]
Alginate	In Tissue engineering, skin regeneration [48]
Keratin	In Tissue engineering, skin regeneration [49]
Starch	Bone regeneration, spinal cord treatment [50]
Elastine	Tissue reconstruction, cell encapsulation [51]
Fibrin	Blood clotting, regenerative medicine [52]

Application of Smart Biopolymers in the Food Industry

Since the last decades, the improvement of biotechnology in the field of the food/food packaging industry has been quite impressive, and biopolymers have played an important role in it. Even though medical science has benefitted more from the recent development in biopolymers, significant growth in the application of biopolymers has also been observed in the food industry. Some of the applications of smart biopolymers in the food industry are discussed below.

Application in the Food Packaging Industry

In modern society, the food industry has to play an important role in terms of the safety and quality of the products manufactured as well as fulfilling the demands of the consumers. The food industry manufactures food with sufficient quality, but they have faced a lot of challenges inmanufacturing food with a longer shelf life [65 - 67]. Over the last few years, biopolymers have emerged as an alternative to smart food packaging material that can prolong the shelf-life of packaging food during storage. These smart biopolymers possess antimicrobial or antioxidant (or both) properties and are used to develop active film or smart packaging material through microencapsulation, which increases the shelf-life of both food, and packaging and also maintains the nutritional value. These smart and intelligent packaging systems can also be utilized in food quality monitoring systems by creating a chemical sensor [68].

Active biopolymer films used in food packaging systems are efficiently prepared by the casting technique at a low cost. In this technique, initially, the biopolymer in the required quantity is suspended in distilled or deionized water to produce the hydrocolloid suspension [69]. The suspension so obtained is then poured into a glass plate and dried till a constant weight is obtained. This technique has been successfully utilized in developing smart and active biopolymer films that can incorporate chitosan, chitosan/starch, starch/glycerol, *etc.* Even though many biopolymers have been used in the development of smart packaging systems, there are only a few biopolymers, such as starch, chitosan, gum, gelatine, *etc.* that exhibit efficiency in packaging systems due to their ability to form network structures as well as thermally stable co-polymers [70].

During the COVID-19 pandemic in early 2020, the lockdown was imposed worldwide, which resulted in a disruption in the supply chain, especially healthy fruit items. As a result, there was a significant loss of fruits in farms due to limited shelf life. During that time, several researchers reported that polyphenol, a biopolymer found in various fruits like apples, grapes, kiwis, *etc.*, can be used as an edible coating on fruits, which helps retain the nutritional quality in fruits during long shelf life [71, 72]. Polyphenol possesses antioxidant, antibacterial, and proliferative properties, and it also improves immune function without any side effects. Examples of polyphenols that are used as edible coatings are quercetin, sinigrin, rhoifolin, hesperetin, herbaceous, *etc* [73 - 76].

Application in Contemporary Food Production

Biopolymers also have applications in contemporary food production, like nutritional fibers, which are traditionally collected from plant sources [77]. It is also observed that nutritional fibers extracted from microbial sources also have health benefits. These nutritional fibers are generally classified as soluble and insoluble fibers based on their solubility in water. Examples of water-soluble fibers are glucan, pectin, and inulin, while the fibers derived from chitosan, cellulose, hemicellulose, and resistant starch are water-insoluble [78, 79]. These fibers are highly helpful for human health as they possess physiological effects in terms of biological response modifiers as well as wound healing benefits. The health benefits of these nutritional fibers are permanent. An important example of nutritional fiber is glucan, which can be derived from various natural sources such as plants, yeast, fungus, bacteria, *etc.* ß-glucan possesses biocompatibility and biodegradability properties which make it useful in various applications [80, 81]. In 2011, the European Commission included glucans as a new food component of functional food [82 - 84].

Application of Polysaccharides in Bakery Products

Starch is a polysaccharide that is second after cellulose in terms of abundance. It is the most abundant carbohydrate reserve in plant tubers and is available as granules. Starch is the major component of bakery food like breads, and cookies [85, 86]. It is also used in the confectionery industry as a gelling and dusting agent. In baking and confectionery industries, starch is used as an alternative to gluten, to substitute fats in chocolates/cookies, and to increase fiber content due to its affordability and non-toxicity. In recent years, the application of modified starches in confectionary, bakery, noodles, and other convenient foods has been increased due to the demand for healthier food [87 - 89].

Other Applications of Smart Polymer

Biopolymer-based smart materials are widely used for application in the biomedical and food industries. However, their application in other fields is limited and not broadly explored. A few other applications of biopolymers in the field of agriculture and energy are discussed below.

Agricultural Application

A number of natural biopolymers in combination with pesticides are commercialized to get controlled release of fungicides and herbicides that help seeds against fungal pathogens [90, 91]. Chitin is a biopolymer that possesses antifungal, antibacterial, antiviral, and antiparasitic properties and can manage plant diseases [92].

Biopolymers for Energy Applications

Since the last few years, applications of biopolymer-based electrolytes have also been studied extensively. When membranes holding ions are dissolved in polymers, then they serve as polymer electrolytes [93]. One example is a carbohydrate polymer-based solid gellan gel electrolyte that can be used as a solar cell material [94, 95].

CONCLUSION AND FUTURE SCOPE

Over the past few decades, the importance of biopolymers for industrial purposes has increased steadily. The increasing interest in biopolymers by the present manufacturers to produce environment-friendly and sustainable goods is due to their renewability, non-toxicity, biodegradability, high mechanical strength, and stiffness. However, its cost-effectiveness should be improved to make it available for all-round uses. In recent times, researchers have emphasized more on smart and active biopolymers due to their wide range of advantages. Smart biopolymers are biocompatible, resilient, non-thrombogenic, highly flexible, easy to shape, and many more. Considering such properties of smart biopolymers, they play an important role in biomedical applications such as modern drug delivery, gene delivery, and tissue engineering. Apart from biomedical applications, these biopolymers are also used in the food industries in terms of smart packaging, and nutritious packaged food (smart or superfood). Smart packaging materials possess antimicrobial as well as antioxidant properties that can prolong the shelf-life of food by maintaining its nutritional value. However, the application of biopolymer-based materials at present is limited to the biomedical science and food industry.

The use of biopolymer-based smart materials in the field of biomedical science, and the food industry has increased by many fold in the last few years. However, its untapped potential in various other fields has yet to be explored/studied. In the future, a large number of biopolymer-based multi-functional materials will be developed for a wide range of applications in various other fields. A lot of future sustainability, and developments in biomedical science, biotechnology, biochemistry, and molecular biology will depend upon the performances of biopolymer-based materials.

REFERENCES

[1] Gopi, S.; Amalraj, A.; Thomas, S. Effective drug delivery system of biopolymers based on nanomaterials and hydrogels-a review. *Drug Des.*, **2016**, *5*(2), 2169-0138. [http://dx.doi.org/10.4172/2169-0138.1000129]

[2] Babu, R.P.; O'Connor, K.; Seeram, R. Current progress on bio-based polymers and their future trends. *Prog. Biomater.*, **2013**, *2*(1), 8.

[http://dx.doi.org/10.1186/2194-0517-2-8] [PMID: 29470779]

[3] Salerno, A.; Pascual, C.D. Bio-based polymers, supercritical fluids and tissue engineering. *Process Biochem.,* **2015**, *50*(5), 826-838.
[http://dx.doi.org/10.1016/j.procbio.2015.02.009]

[4] Yates, M.R.; Barlow, C.Y. Life cycle assessments of biodegradable, commercial biopolymers—A critical review. *Resour. Conserv. Recycling,* **2013**, *78*, 54-66.
[http://dx.doi.org/10.1016/j.resconrec.2013.06.010]

[5] Van de Velde, K.; Kiekens, P. Biopolymers: overview of several properties and consequences on their applications. *Polym. Test.,* **2002**, *21*(4), 433-442.
[http://dx.doi.org/10.1016/S0142-9418(01)00107-6]

[6] Kim, B.S.; Baez, C.E.; Atala, A. Biomaterials for tissue engineering. *World J. Urol.,* **2000**, *18*(1), 2-9.
[http://dx.doi.org/10.1007/s003450050002] [PMID: 10766037]

[7] Galaev, I.; Mattiasson, B. 'Smart' polymers and what they could do in biotechnology and medicine. *Trends Biotechnol.,* **1999**, *17*(8), 335-340.
[http://dx.doi.org/10.1016/S0167-7799(99)01345-1] [PMID: 10407406]

[8] Dušek, K.; Patterson, D. Transition in swollen polymer networks induced by intramolecular condensation. *J Polym Sci A-2,* **1968**, *6*(7), 1209-1216.

[9] Dušek, K.; Prins, W. *Structure and elasticity of non-crystalline polymer networks*; In Fortschritte der Hochpolymeren-Forschung: Springer Berlin Heidelberg, Berlin, Heidelberg, **2006**, pp. 1-102.

[10] Hoffman, A.S. "Intelligent" polymers in medicine and biotechnology. *Artif. Organs,* **1995**, *19*(5), 458-467.
[http://dx.doi.org/10.1111/j.1525-1594.1995.tb02359.x] [PMID: 7625927]

[11] Osada, Y.; Ross-Murphy, S.B. Intelligent Gels. *Sci. Am.,* **1993**, *268*(5), 82-87.
[http://dx.doi.org/10.1038/scientificamerican0593-82]

[12] Urry, D.W. Five axioms for the functional design of peptide-based polymers as molecular machines and materials: Principle for macromolecular assemblies. *Biopolymers,* **1998**, *47*(2), 167-178.
[http://dx.doi.org/10.1002/(SICI)1097-0282(1998)47:2<167::AID-BIP6>3.0.CO;2-S]

[13] Barker, S.L.R.; Ross, D.; Tarlov, M.J.; Gaitan, M.; Locascio, L.E. Control of flow direction in microfluidic devices with polyelectrolyte multilayers. *Anal. Chem.,* **2000**, *72*(24), 5925-5929.
[http://dx.doi.org/10.1021/ac0008690] [PMID: 11140758]

[14] Kikuchi, A.; Okano, T. Pulsatile drug release control using hydrogels. *Adv. Drug Deliv. Rev.,* **2002**, *54*(1), 53-77.
[http://dx.doi.org/10.1016/S0169-409X(01)00243-5] [PMID: 11755706]

[15] Li, S.K.; D'Emanuele, A. On–off transport through a thermoresponsive hydrogel composite membrane. *J. Control. Release,* **2001**, *75*(1-2), 55-67.
[http://dx.doi.org/10.1016/S0168-3659(01)00365-0] [PMID: 11451497]

[16] Kost, J.; Langer, R. Responsive polymeric delivery systems. *Adv. Drug Deliv. Rev.,* **2012**, *64*, 327-341.
[http://dx.doi.org/10.1016/j.addr.2012.09.014] [PMID: 11259837]

[17] Peelman, N.; Ragaert, P.; De Meulenaer, B.; Adons, D.; Peeters, R.; Cardon, L.; Van Impe, F.; Devlieghere, F. Application of bioplastics for food packaging. *Trends Food Sci. Technol.,* **2013**, *32*(2), 128-141.
[http://dx.doi.org/10.1016/j.tifs.2013.06.003]

[18] Baranwal, J.; Barse, B.; Fais, A.; Delogu, G.L.; Kumar, A. Biopolymer: A sustainable material for food and medical applications. *Polymers (Basel),* **2022**, *14*(5), 983.
[http://dx.doi.org/10.3390/polym14050983] [PMID: 35267803]

[19] Liodakis, S.; Bakirtzis, D.; Dimitrakopoulos, A. Ignition characteristics of forest species in relation to

thermal analysis data. *Thermochim. Acta,* **2002**, *390*(1-2), 83-91. [http://dx.doi.org/10.1016/S0040-6031(02)00077-1]

[20] Kaplan, D.L. *Introduction to biopolymers from renewable resources*; Biopolymers from renewable resources: Springer Berlin Heidelberg, Berlin, Heidelberg, **1998**, pp. 1-29. [http://dx.doi.org/10.1007/978-3-662-03680-8_1]

[21] Kaplan, D.L.; Mayer, J.M.; Ball, D.; McCassie, J.; Allen, A.L.; Stenhouse, P. Fundamentals of biodegradable polymers. *Biodegradable polymers and packaging,* **1993**, *8*(5), 1-42.

[22] Kumar, M.N.V.R.; Muzzarelli, R.A.A.; Muzzarelli, C.; Sashiwa, H.; Domb, A.J. Chitosan chemistry and pharmaceutical perspectives. *Chem. Rev.,* **2004**, *104*(12), 6017-6084. [http://dx.doi.org/10.1021/cr030441b] [PMID: 15584695]

[23] Rouilly, A.; Rigal, L. Agro-materials: a bibliographic review. *Journal of J. Macomol. Sci.-Part C. Polym. Rev. (Phila. Pa.),* **2002**, *42*(4), 441-479.

[24] Chandra, R.; Rustgi, R. Biodegradable polymers. *Prog. Polym. Sci.,* **1998**, *23*(7), 1273-1335. [http://dx.doi.org/10.1016/S0079-6700(97)00039-7]

[25] Klemm, D.; Heublein, B.; Fink, H.P.; Bohn, A. Cellulose: fascinating biopolymer and sustainable raw material. *Angew. Chem. Int. Ed.,* **2005**, *44*(22), 3358-3393. [http://dx.doi.org/10.1002/anie.200460587] [PMID: 15861454]

[26] Kurokawa, K.; Yamashita, K.; Doi, Y.; Abe, H. Structural effects of terminal groups on nonenzymatic and enzymatic degradations of end-capped poly(L-lactide). *Biomacromolecules,* **2008**, *9*(3), 1071-1078. [http://dx.doi.org/10.1021/bm701259r] [PMID: 18275150]

[27] Leoni, E.; Cancellieri, D.; Balbi, N.; Tomi, P.; Bernardini, A.F.; Kaloustian, J.; Marcelli, T. Thermal degradation of pinus pinaster needles by dsc, part 2: Kinetics of exothermic phenomena. *J. Fire Sci.,* **2003**, *21*(2), 117-130. [http://dx.doi.org/10.1177/0734904103021002002]

[28] Branca, C.; Di Blasi, C. Kinetics of the isothermal degradation of wood in the temperature range 528–708 K. *J. Anal. Appl. Pyrolysis,* **2003**, *67*(2), 207-219. [http://dx.doi.org/10.1016/S0165-2370(02)00062-1]

[29] Pattanashetti, N.A.; Heggannavar, G.B.; Kariduraganavar, M.Y. Smart biopolymers and their biomedical applications. *Procedia Manuf.,* **2017**, *12*, 263-279. [http://dx.doi.org/10.1016/j.promfg.2017.08.030]

[30] Schmaljohann, D. Thermo- and pH-responsive polymers in drug delivery. *Adv. Drug Deliv. Rev.,* **2006**, *58*(15), 1655-1670. [http://dx.doi.org/10.1016/j.addr.2006.09.020] [PMID: 17125884]

[31] Bawa, P.; Pillay, V.; Choonara, Y.E.; du Toit, L.C. Stimuli-responsive polymers and their applications in drug delivery. *Biomed. Mater.,* **2009**, *4*(2), 022001. [http://dx.doi.org/10.1088/1748-6041/4/2/022001] [PMID: 19261988]

[32] Nakayama, M.; Okano, T.; Miyazaki, T.; Kohori, F.; Sakai, K.; Yokoyama, M. Molecular design of biodegradable polymeric micelles for temperature-responsive drug release. *J. Control. Release,* **2006**, *115*(1), 46-56. [http://dx.doi.org/10.1016/j.jconrel.2006.07.007] [PMID: 16920217]

[33] Ward, M.A.; Georgiou, T.K. Thermoresponsive polymers for biomedical applications. *Polymers (Basel),* **2011**, *3*(3), 1215-1242. [http://dx.doi.org/10.3390/polym3031215]

[34] Mastalska-Popławska, J.; Sikora, M.; Izak, P.; Góral, Z. Applications of starch and its derivatives in bioceramics. *J. Biomater. Appl.,* **2019**, *34*(1), 12-24. [http://dx.doi.org/10.1177/0885328219844972] [PMID: 31014208]

[35] K.; Xuejun, C.; Jinan, X.; Hur, B.K. Synthesis and application of a light-sensitive polymer forming aqueous two-phase systems. *J. Ind. Eng. Chem.*, **2007**, *13*(3), 424-428.

[36] Shiga, T. Deformation and viscoelastic behavior of polymer gels in electric fields. *Adv. Polym. Sci.*, **1997**, *134*, 131-163.
[http://dx.doi.org/10.1007/3-540-68449-2_2]

[37] Jane, J. Starch properties, modifications, and applications. *J. Macromol. Sci. Part A Pure Appl. Chem.*, **1995**, *32*(4), 751-757.
[http://dx.doi.org/10.1080/10601329508010286]

[38] Neuendorf, R.E.; Saiz, E.; Tomsia, A.P.; Ritchie, R.O. Adhesion between biodegradable polymers and hydroxyapatite: Relevance to synthetic bone-like materials and tissue engineering scaffolds. *Acta Biomater.*, **2008**, *4*(5), 1288-1296.
[http://dx.doi.org/10.1016/j.actbio.2008.04.006] [PMID: 18485842]

[39] Chen, Y.; Hung, S.T.; Chou, E.; Wu, H.S. Review of polyhydroxy alkanoates materials and other biopolymers for medical applications. *Mini Rev. Org. Chem.*, **2018**, *15*(2), 105-121.
[http://dx.doi.org/10.2174/1570193X14666170721153036]

[40] Tsai, C.H.; Wang, P.Y.; Lin, I.C.; Huang, H.; Liu, G.S.; Tseng, C.L. Ocular drug delivery: Role of degradable polymeric nanocarriers for ophthalmic application. *Int. J. Mol. Sci.*, **2018**, *19*(9), 2830.
[http://dx.doi.org/10.3390/ijms19092830] [PMID: 30235809]

[41] Yahya, E.B.; Amirul, A.A.; H P S, A.K.; Olaiya, N.G.; Iqbal, M.O.; Jummaat, F.; A K, A.S.; Adnan, A.S. HPS, A.K.; Olaiya, N.G.; Iqbal, M.O.; Jummaat, F.; Adnan, A.S. Insights into the role of biopolymer aerogel scaffolds in tissue engineering and regenerative medicine. *Polymers (Basel)*, **2021**, *13*(10), 1612.
[http://dx.doi.org/10.3390/polym13101612] [PMID: 34067569]

[42] Chaikof, E.L.; Matthew, H.; Kohn, J.; Mikos, A.G.; Prestwich, G.D.; Yip, C.M. Biomaterials and scaffolds in reparative medicine. *Ann. N. Y. Acad. Sci.*, **2002**, *961*(1), 96-105.
[http://dx.doi.org/10.1111/j.1749-6632.2002.tb03057.x] [PMID: 12081873]

[43] Hook, A.L.; Anderson, D.G.; Langer, R.; Williams, P.; Davies, M.C.; Alexander, M.R. High throughput methods applied in biomaterial development and discovery. *Biomaterials*, **2010**, *31*(2), 187-198.
[http://dx.doi.org/10.1016/j.biomaterials.2009.09.037] [PMID: 19815273]

[44] Reddy, N.; Reddy, R.; Jiang, Q. Crosslinking biopolymers for biomedical applications. *Trends Biotechnol.*, **2015**, *33*(6), 362-369.
[http://dx.doi.org/10.1016/j.tibtech.2015.03.008] [PMID: 25887334]

[45] Wróblewska-Krepsztul, J.; Rydzkowski, T.; Michalska-Pożoga, I.; Thakur, V.K. Biopolymers for biomedical and pharmaceutical applications: Recent advances and overview of alginate electrospinning. *Nanomaterials (Basel)*, **2019**, *9*(3), 404.
[http://dx.doi.org/10.3390/nano9030404] [PMID: 30857370]

[46] Gagliardi, A.; Giuliano, E.; Venkateswararao, E.; Fresta, M.; Bulotta, S.; Awasthi, V.; Cosco, D. Biodegradable polymeric nanoparticles for drug delivery to solid tumors. *Front. Pharmacol.*, **2021**, *12*, 601626.
[http://dx.doi.org/10.3389/fphar.2021.601626] [PMID: 33613290]

[47] Tripathi, D.; Rastogi, K.; Tyagi, P.; Rawat, H.; Mittal, G.; Jamini, A.; Singh, H.; Tyagi, A. Comparative analysis of collagen and chitosan-based dressing for haemostatic and wound healing application. *AAPS PharmSciTech*, **2021**, *22*(3), 76.
[http://dx.doi.org/10.1208/s12249-021-01944-9] [PMID: 33595780]

[48] Álvarez-Castillo, E.; Aguilar, J.M.; Bengoechea, C.; López-Castejón, M.L.; Guerrero, A. Rheology and water absorption properties of alginate–soy protein composites. *Polymers (Basel)*, **2021**, *13*(11), 1807.

[http://dx.doi.org/10.3390/polym13111807] [PMID: 34072653]

[49] Goswami, M.; Rekhi, P.; Debnath, M.; Ramakrishna, S. Microbial polyhydroxy alkanoates granules: an approach targeting biopolymer for medical applications and developing bone scaffolds. *Molecules,* **2021**, *26*(4), 860.
[http://dx.doi.org/10.3390/molecules26040860] [PMID: 33562111]

[50] Sarangthem, V.; Singh, T.D.; Dinda, A.K. Emerging role of elastin-like polypeptides in regenerative medicine. *Adv. Wound Care (New Rochelle),* **2021**, *10*(5), 257-269.
[http://dx.doi.org/10.1089/wound.2019.1085] [PMID: 32602815]

[51] Sellappan, L.K.; Anandhavelu, S.; Doble, M.; Perumal, G.; Jeon, J.H.; Vikraman, D.; Kim, H.S. Biopolymer film fabrication for skin mimetic tissue regenerative wound dressing applications. *Int. J. Polym. Mater.,* **2022**, *71*(3), 196-207.
[http://dx.doi.org/10.1080/00914037.2020.1817019]

[52] Nelson, D.W.; Gilbert, R.J. Extracellular matrix-mimetic hydrogels for treating neural tissue injury: a focus on fibrin, hyaluronic acid, and elastin-like polypeptide hydrogels. *Adv. Healthc. Mater.,* **2021**, *10*(22), 2101329.
[http://dx.doi.org/10.1002/adhm.202101329] [PMID: 34494398]

[53] Rebelo, R.; Fernandes, M.; Fangueiro, R. Biopolymers in medical implants: a brief review. *Procedia Eng.,* **2017**, *200*, 236-243.
[http://dx.doi.org/10.1016/j.proeng.2017.07.034]

[54] Nitta, S.; Numata, K. Biopolymer-based nanoparticles for drug/gene delivery and tissue engineering. *Int. J. Mol. Sci.,* **2013**, *14*(1), 1629-1654.
[http://dx.doi.org/10.3390/ijms14011629] [PMID: 23344060]

[55] Nagarajan, S.; Radhakrishnan, S.; Kalkura, S.N.; Balme, S.; Miele, P.; Bechelany, M. Overview of protein-based biopolymers for biomedical application. *Macromol. Chem. Phys.,* **2019**, *220*(14), 1900126.
[http://dx.doi.org/10.1002/macp.201900126]

[56] Nagarajan, S.; Abessolo Ondo, D.; Gassara, S.; Bechelany, M.; Balme, S.; Miele, P.; Kalkura, N.; Pochat-Bohatier, C. Porous gelatin membrane obtained from pickering emulsions stabilized by graphene oxide. *Langmuir,* **2018**, *34*(4), 1542-1549.
[http://dx.doi.org/10.1021/acs.langmuir.7b03426] [PMID: 29278504]

[57] Kharkar, P.M.; Kiick, K.L.; Kloxin, A.M. Designing degradable hydrogels for orthogonal control of cell microenvironments. *Chem. Soc. Rev.,* **2013**, *42*(17), 7335-7372.
[http://dx.doi.org/10.1039/C3CS60040H] [PMID: 23609001]

[58] Sridhar, R.; Lakshminarayanan, R.; Madhaiyan, K.; Amutha Barathi, V.; Lim, K.H.C.; Ramakrishna, S. Electrosprayed nanoparticles and electrospun nanofibers based on natural materials: applications in tissue regeneration, drug delivery and pharmaceuticals. *Chem. Soc. Rev.,* **2015**, *44*(3), 790-814.
[http://dx.doi.org/10.1039/C4CS00226A] [PMID: 25408245]

[59] Chigurupati, N.; Saiki, L.; Gayser, C., Jr; Dash, A.K. Evaluation of red cabbage dye as a potential natural color for pharmaceutical use. *Int. J. Pharm.,* **2002**, *241*(2), 293-299.
[http://dx.doi.org/10.1016/S0378-5173(02)00246-6] [PMID: 12100856]

[60] Thandapani, G.; P, S.P.; P N, S.; Sukumaran, A. Size optimization and *in vitro* biocompatibility studies of chitosan nanoparticles. *Int. J. Biol. Macromol.,* **2017**, *104*(Pt B), 1794-1806.
[http://dx.doi.org/10.1016/j.ijbiomac.2017.08.057] [PMID: 28807691]

[61] Ahmed, H.; Hashim, A. Structure, optical, electronic and chemical characteristics of novel (PVA-CoO) structure doped with silicon carbide. *Silicon,* **2021**, *13*(12), 4331-4344.
[http://dx.doi.org/10.1007/s12633-020-00723-8]

[62] H P S, A.K.; Saurabh, C.K.; A S, A.; Nurul Fazita, M.R.; Syakir, M.I.; Davoudpour, Y.; Rafatullah, M.; Abdullah, C.K.; M Haafiz, M.K.; Dungani, R. A review on chitosan-cellulose blends and

nanocellulose reinforced chitosan biocomposites: Properties and their applications. *Carbohydr. Polym.,* **2016**, *150*, 216-226.
[http://dx.doi.org/10.1016/j.carbpol.2016.05.028] [PMID: 27312632]

[63] Sharifianjazi, F.; Khaksar, S.; Esmaeilkhanian, A.; Bazli, L.; Eskandarinezhad, S.; Salahshour, P.; Sadeghi, F.; Rostamnia, S.; Vahdat, S.M. Advancements in fabrication and application of chitosan composites in implants and dentistry: A review. *Biomolecules,* **2022**, *12*(2), 155.
[http://dx.doi.org/10.3390/biom12020155] [PMID: 35204654]

[64] Su, S.; Kang, P.M. Systemic review of biodegradable nanomaterials in nanomedicine. *Nanomaterials (Basel),* **2020**, *10*(4), 656.
[http://dx.doi.org/10.3390/nano10040656] [PMID: 32244653]

[65] Laine, C.; Harlin, A.; Hartman, J.; Hyvärinen, S.; Kammiovirta, K.; Krogerus, B.; Pajari, H.; Rautkoski, H.; Setälä, H.; Sievänen, J.; Uotila, J.; Vähä-Nissi, M. Hydroxyalkylated xylans – Their synthesis and application in coatings for packaging and paper. *Ind. Crops Prod.,* **2013**, *44*, 692-704.
[http://dx.doi.org/10.1016/j.indcrop.2012.08.033]

[66] Moraga, G.; Talens, P.; Moraga, M.J.; Martínez-Navarrete, N. Implication of water activity and glass transition on the mechanical and optical properties of freeze-dried apple and banana slices. *J. Food Eng.,* **2011**, *106*(3), 212-219.
[http://dx.doi.org/10.1016/j.jfoodeng.2011.05.009]

[67] Rutz, J.K.; Zambiazi, R.C.; Borges, C.D.; Krumreich, F.D.; da Luz, S.R.; Hartwig, N.; da Rosa, C.G. Microencapsulation of purple Brazilian cherry juice in xanthan, tara gums and xanthan-tara hydrogel matrixes. *Carbohydr. Polym.,* **2013**, *98*(2), 1256-1265.
[http://dx.doi.org/10.1016/j.carbpol.2013.07.058] [PMID: 24053801]

[68] Bankura, K.P.; Maity, D.; Mollick, M.M.R.; Mondal, D.; Bhowmick, B.; Bain, M.K.; Chakraborty, A.; Sarkar, J.; Acharya, K.; Chattopadhyay, D. Synthesis, characterization and antimicrobial activity of dextran stabilized silver nanoparticles in aqueous medium. *Carbohydr. Polym.,* **2012**, *89*(4), 1159-1165.
[http://dx.doi.org/10.1016/j.carbpol.2012.03.089] [PMID: 24750927]

[69] Islam, M.S.; Yeum, J.H. Electrospun pullulan/poly(vinyl alcohol)/silver hybrid nanofibers: Preparation and property characterization for antibacterial activity. *Colloids Surf. A Physicochem. Eng. Asp.,* **2013**, *436*, 279-286.
[http://dx.doi.org/10.1016/j.colsurfa.2013.07.001]

[70] Venkatesan, J.; Singh, S.K.; Anil, S.; Kim, S.K.; Shim, M.S. Preparation, characterization and biological applications of biosynthesized silver nanoparticles with chitosan-fucoidan coating. *Molecules,* **2018**, *23*(6), 1429.
[http://dx.doi.org/10.3390/molecules23061429] [PMID: 29895803]

[71] Chitrakar, B.; Zhang, M.; Bhandari, B. Improvement strategies of food supply chain through novel food processing technologies during COVID-19 pandemic. *Food Control,* **2021**, *125*, 108010.
[http://dx.doi.org/10.1016/j.foodcont.2021.108010] [PMID: 33679006]

[72] Ghosh, M.; Singh, A.K. Potential of engineered nanostructured biopolymer based coatings for perishable fruits with Coronavirus safety perspectives. *Prog. Org. Coat.,* **2022**, *163*, 106632.
[http://dx.doi.org/10.1016/j.porgcoat.2021.106632] [PMID: 34931104]

[73] Mehany, T.; Khalifa, I.; Barakat, H.; Althwab, S.A.; Alharbi, Y.M.; El-Sohaimy, S. Polyphenols as promising biologically active substances for preventing SARS-CoV-2: A review with research evidence and underlying mechanisms. *Food Biosci.,* **2021**, *40*, 100891.
[http://dx.doi.org/10.1016/j.fbio.2021.100891] [PMID: 33495727]

[74] Paraiso, I.L.; Revel, J.S.; Stevens, J.F. Potential use of polyphenols in the battle against COVID-19. *Curr. Opin. Food Sci.,* **2020**, *32*, 149-155.
[http://dx.doi.org/10.1016/j.cofs.2020.08.004] [PMID: 32923374]

[75] Tongnuanchan, P.; Benjakul, S.; Prodpran, T. Physico-chemical properties, morphology and

antioxidant activity of film from fish skin gelatin incorporated with root essential oils. *J. Food Eng.,* **2013**, *117*(3), 350-360.
[http://dx.doi.org/10.1016/j.jfoodeng.2013.03.005]

[76] Kreyenschmidt, J.; Christiansen, H.; Hübner, A.; Raab, V.; Petersen, B. A novel photochromic time–temperature indicator to support cold chain management. *Int. J. Food Sci. Technol.,* **2010**, *45*(2), 208-215.
[http://dx.doi.org/10.1111/j.1365-2621.2009.02123.x]

[77] Peltzer, M.; Delgado, J.F.; Salvay, A.G.; Wagner, J.R. β-Glucan, a promising polysaccharide for bio-based films developments for food contact materials and medical applications. *Curr. Org. Chem.,* **2018**, *22*(12), 1249-1254.
[http://dx.doi.org/10.2174/1385272822666171129153633]

[78] Felton, L.A. Mechanisms of polymeric film formation. *Int. J. Pharm.,* **2013**, *457*(2), 423-427.
[http://dx.doi.org/10.1016/j.ijpharm.2012.12.027] [PMID: 23305867]

[79] Thammakiti, S.; Suphantharika, M.; Phaesuwan, T.; Verduyn, C. Preparation of spent brewer's yeast β-glucans for potential applications in the food industry. *Int. J. Food Sci. Technol.,* **2004**, *39*(1), 21-29.
[http://dx.doi.org/10.1111/j.1365-2621.2004.00742.x]

[80] Vetvicka, V.; Vetvickova, J. β(1-3)-D-glucan affects adipogenesis, wound healing and inflammation. *Orient. Pharm. Exp. Med.,* **2011**, *11*(3), 169-175.
[http://dx.doi.org/10.1007/s13596-011-0024-4]

[81] Zhu, F.; Du, B.; Xu, B. A critical review on production and industrial applications of beta-glucans. *Food Hydrocoll.,* **2016**, *52*, 275-288.
[http://dx.doi.org/10.1016/j.foodhyd.2015.07.003]

[82] Henrion, M.; Francey, C.; Lê, K.A.; Lamothe, L. Cereal B-glucans: The impact of processing and how it affects physiological responses. *Nutrients,* **2019**, *11*(8), 1729.
[http://dx.doi.org/10.3390/nu11081729] [PMID: 31357461]

[83] Levi, S.; Rac, V.; Manojlovi, V.; Raki, V.; Bugarski, B.; Flock, T.; Krzyczmonik, K.E.; Nedovi, V. Limonene encapsulation in alginate/poly (vinyl alcohol). *Procedia Food Sci.,* **2011**, *1*, 1816-1820.
[http://dx.doi.org/10.1016/j.profoo.2011.09.266]

[84] Ramos, M.; Beltrán, A.; Peltzer, M.; Valente, A.J.M.; Garrigós, M.C. Release and antioxidant activity of carvacrol and thymol from polypropylene active packaging films. *Lebensm. Wiss. Technol.,* **2014**, *58*(2), 470-477.
[http://dx.doi.org/10.1016/j.lwt.2014.04.019]

[85] Ludwicka, K.; Kaczmarek, M.; Białkowska, A. Bacterial nanocellulose—A biobased polymer for active and intelligent food packaging applications: Recent advances and developments. *Polymers (Basel),* **2020**, *12*(10), 2209.
[http://dx.doi.org/10.3390/polym12102209] [PMID: 32993082]

[86] Eerlingen, R.C.; Delcour, J.A. Formation, analysis, structure and properties of type III enzyme resistant starch. *J. Cereal Sci.,* **1995**, *22*(2), 129-138.
[http://dx.doi.org/10.1016/0733-5210(95)90042-X]

[87] Chen, S.; Chen, Y.; Wang, Z.; Chen, H.; Fan, D. Renewable bio-based adhesive fabricated from a novel biopolymer and soy protein. *RSC Advances,* **2021**, *11*(19), 11724-11731.
[http://dx.doi.org/10.1039/D1RA00766A] [PMID: 35423652]

[88] Wilderjans, E.; Luyts, A.; Goesaert, H.; Brijs, K.; Delcour, J.A. A model approach to starch and protein functionality in a pound cake system. *Food Chem.,* **2010**, *120*(1), 44-51.
[http://dx.doi.org/10.1016/j.foodchem.2009.09.067]

[89] Williams, P.A.; Phillips, G.O. *Introduction to food hydrocolloids*; Woodhead publishing, **2021**, pp. 3-26.
[http://dx.doi.org/10.1016/B978-0-12-820104-6.00017-6]

[90] Munkvold, G.P. Seed pathology progress in academia and industry. *Annu. Rev. Phytopathol.,* **2009**, *47*(1), 285-311.
[http://dx.doi.org/10.1146/annurev-phyto-080508-081916] [PMID: 19400648]

[91] Avérous, L. Biodegradable multiphase systems based on plasticized starch: a review. *J. Macromol. Sci. Part C Polym. Rev.,* **2004**, *44*(3), 231-274.
[http://dx.doi.org/10.1081/MC-200029326]

[92] Akelah, A. Applications of functionalized polymers in agriculture. *J. Islam. Acad. Sci.,* **1990**, *3*(1), 49-61.

[93] Park, C.; Kim, D.W.; Prakash, J.; Sun, Y.K. Electrochemical stability and conductivity enhancement of composite polymer electrolytes. *Solid State Ion.,* **2003**, *159*(1-2), 111-119.
[http://dx.doi.org/10.1016/S0167-2738(03)00025-0]

[94] Shuhaimi, N.E.A.; Teo, L.P.; Woo, H.J.; Majid, S.R.; Arof, A.K. Electrical double-layer capacitors with plasticized polymer electrolyte based on methyl cellulose. *Polym. Bull.,* **2012**, *69*(7), 807-826.
[http://dx.doi.org/10.1007/s00289-012-0763-5]

[95] Singh, R.; Bhattacharya, B.; Rhee, H.W.; Singh, P.K. Solid gellan gum polymer electrolyte for energy application. *Int. J. Hydrogen Energy,* **2015**, *40*(30), 9365-9372.
[http://dx.doi.org/10.1016/j.ijhydene.2015.05.084]

CHAPTER 10

Synthesis and Applications of Smart Polymer Nanocomposites in the Flexible Electronic Sensing

Basanta Saikia[1,*] and **Champak Dutta**[2]

[1] *Department of Chemistry, KU Leuven, Leuven-3001, Belgium*

[2] *Department of Chemistry, Bahona College, Assam-785101, India*

Abstract: The use of smart electronic devices incorporated with sensors has become an indispensable part of the modern lifestyle. Such devices utilize the concept of flexible electronic sensing. The growing demand for flexible sensors is understandable because of their versatile application in wearable devices and other fields like robotics in the health sector. Continuous healthcare monitoring and personalized health concerns necessitate the development of flexible wearable devices. Flexible sensors play a crucial role in fabricating such wearable devices. In the realm of this, polymer nanocomposites (PNCs) have emerged as primary materials in flexible sensor design owing to their unique combination of electrical and mechanical properties *viz.* conductivity, mechanical properties, and chemical resistance. This book chapter explores the role of polymer nanocomposites in flexible sensor development. Additionally, various fabrication methods for sensor design with PNCs have been discussed. Moreover, recent advancements in utilizing PNCs for various sensor types are highlighted, paving the way for a new generation of comfortable and reliable wearable health monitoring devices.

Keywords: Core-shell structure, Diffusion-controlled, Drug delivery system, Flexible electronic, Magnetic hyperthermia, Magnetically-guided delivery, Mult stimuli-responsive, pH-responsive, Polymer-drug carriers, Smart polymers, Solvent-activated, Superparamagnetic iron oxide nanoparticles, Temperature responsive.

INTRODUCTION

In the last two decades, there has been rapid development in sensor-based technology owing to its versatile application in different areas [1 - 7]. In general, sensors can be classified into two broad categories based on their mechanical properties, *i.e.* rigid and flexible. Rigid sensors are usually made of low-cost

* **Corresponding author Basanta Saikia:** Department of Chemistry, KU Leuven, Leuven-3001, Belgium; E-mail: bsaikia1@gmail.com

Partha Pratim Saikia & Bhaskar Jyoti Saikia (Eds.)

materials, but they have limited flexibility. Due to their lack of flexibility, the application of rigid sensors is also quite limited. However, sensors for continuous health monitoring and other specific uses are required to be continually bent, stretched, or put under pressure, with a larger extent of flexibility [3, 7]. In this context, polymer nanocomposites (PNCs) have presented themselves as an excellent option for fabricating such sensors [7, 8]. PNCs are heterogeneous (multiphase) polymeric materials composed of at least one nano-scale phase (known as nanofiller) that is dispersed in a second phase (known as matrix) [9]. This heterogeneous blending provides PNCs with the combination of the individual properties of their constituents [9 - 11]. Owing to their excellent physio-chemical properties, they have been widely explored since their discovery in the early 1990s. The inherent flexible property of PNCs also allows for the creation of complex shapes, which can lead to reductions in the size, weight, and cost of the fabricated devices [7]. At present time, advancements in sensor technology have been achieved to develop flexible and biocompatible sensors based on PNCs [7, 8, 12, 13]. Such sensors can be integrated into wearable films, patches, and even temporary tattoos, *etc*. These devices utilise the concept of electronic sensing [14, 15]. Electronic sensing usually involves a sensor that responds to changes in the property being measured and converts it into an electrical signal [14]. The signal received can further be processed by suitable electronic circuits to interpret the data and provide an output that can be perceived by the users, *via* digital displays, analogue meters, or digital communication interfaces. The sensors continuously collect physiological data in real time, allowing for remote patient management by healthcare providers [1, 3, 16]. Machine learning algorithms enable healthcare professionals to optimize treatment based on the collected data from the sensors, assessing medication efficacy, safety, and recovery progress. These sensors are often referred to as electronic skin and are designed to mimic skin characteristics [3]. They can be placed directly on the human body to measure vital signs such as body temperature, heartbeat, and sweat composition. Advances in materials science and electronics have allowed for the development of electronic skin sensors with high resolution and fast response time while being flexible and stretchable [3]. Their flexibility makes them easier to integrate onto curved surfaces like the neck and wrist.

Considering the various applications of PNCs for the development of flexible sensors, in this chapter, we have highlighted the recent development of various types of flexible sensors. Various techniques used for the fabrication of flexible sensors with PNCs have also been discussed. With the advancement of material science and technology, sensors with enhanced flexibility, comfort, and responses are anticipated. Various applications of wearable sensors in the healthcare sector have been illustrated in Fig. (**1**) below.

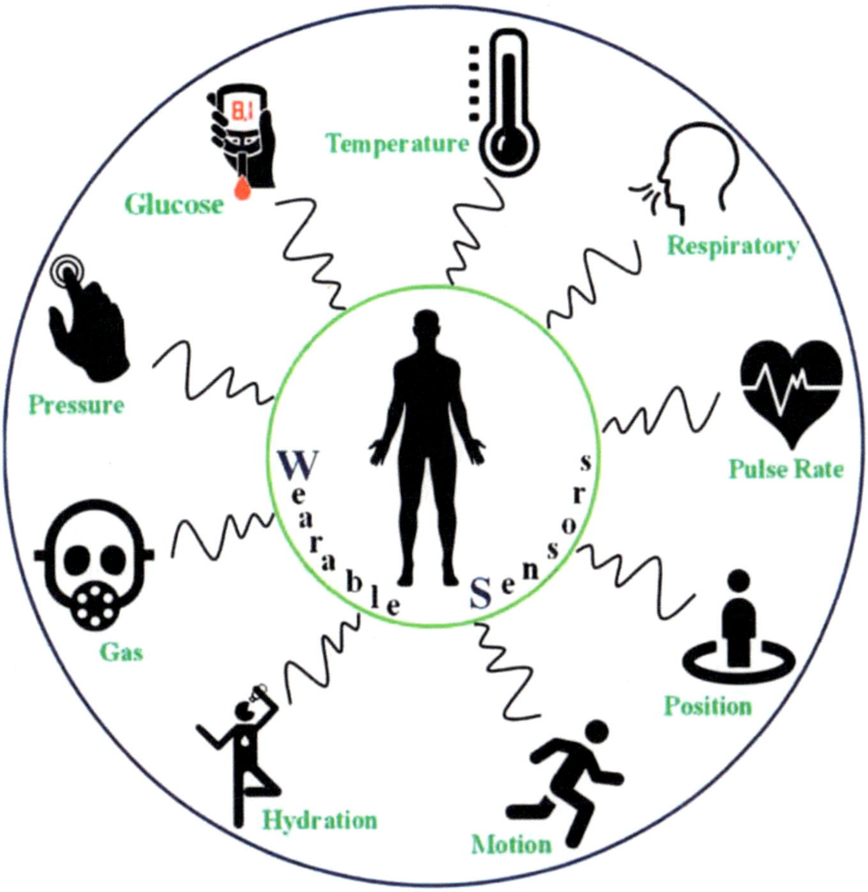

Fig. (1). Schematic of representative wearable healthcare sensing devices [16].

Material and Device Fabrication

The sensor's performance is influenced by the technique used for its fabrication [6, 17, 18]. Depending on the required resolution, sensor size and cost, flexible sensors are processed using various methods. Therefore, appropriate fabrication techniques are necessary for specific applications. For example, the Aerosol jet printing method is used for the fabrication of nano/microsensors at a very high resolution [6]. This is a 3D printing method which has been widely adopted for patterning high-resolution nanomaterial, thin film conductors, semiconductors, and dielectrics. However, as this method is not cost-effective, inkjet printing is used to print similar sensors with a lower resolution [6]. The inkjet printing technique is preferred for the large-scale fabrication of flexible electronic sensors. It is cost-effective because of its low material waste [19]. As a cost-effective and precise manufacturing method, inkjet printing is a versatile, controllable, and automatic computer-aided approach that uses drop-on-demand technology. In this technique, it is possible to customize sensor patterns, by precisely controlling the placement of ink droplets. Inkjet-printed sensors are advantageous considering their good flexibility, lightweight and ease of processing. Various electronic sensors have been fabricated using the inkjet printing method as compared to conventional manufacturing approaches. Selecting a fabrication technique for flexible sensors often requires balancing cost, resolution and usability. Alternative methods for the development of flexible sensors include photolithography, mechanical mixing, flash freezing and vacuum drying, printing and dip-coating, screen printing and air spray coating, drop casting, spin coating *etc*. In this context, soft lithography is a non-destructive fabrication of nanostructures for designing flexible sensors. Thermal nano-imprint lithography (NIL) has been used by Tang *et al.* for the direct fabrication of sub-100 nm nanowires on a flexible PDMS substrate for sensing device integration [20]. This nanowire-based NH_3 sensor with a detection limit of 100 parts per billion, exhibits robust flexibility, mechanical durability and power consumption as low as 3 μW. Wang *et al.* reported the photolithography-based fabrication of a flexible graphene-based Hall sensor [21]. In a recent review, various fabrication technologies, their resolution limit and related device performance, sensitivity, detection limit, reproducibility, and robustness have been discussed by Wong *et al.* [22]. Fig. (**2**) below illustrates the relation between resolution (nm) and throughput (cm^2h^{-1}) of various fabrication techniques.

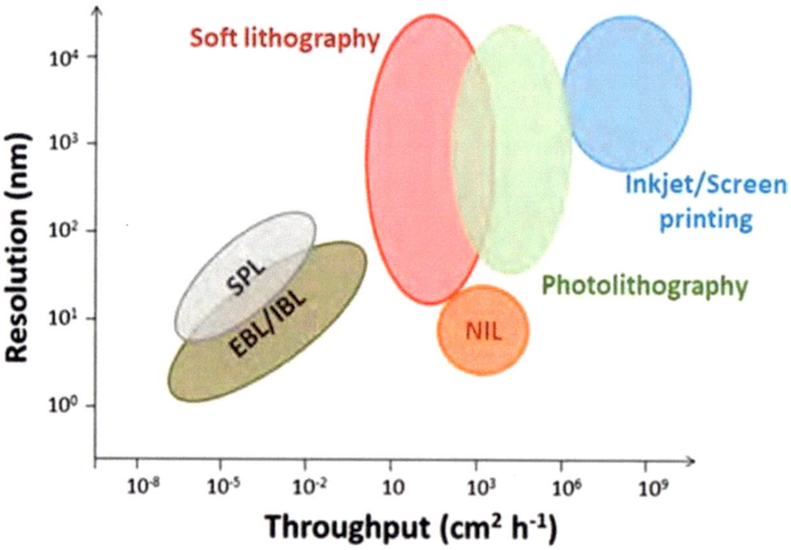

Fig. (2). Graph of feature resolution *versus* fabrication throughput showing the relative location of various lithographic and printing processes [22].

Strain, pressure, and temperature flexible sensors are crucial components in wearable flexible devices [7]. There is a constant effort by researchers to improve these sensors' quality by focusing on the materials used. Nanomaterials like graphene [14], carbon nanotubes [23], and MXene [24] have been used to develop futuristic flexible sensors. Furthermore, composite materials (*e.g.*, CNT–Ag nanoparticle (AgNP) composites *etc.*) [25], cellulose nanofibrils [26], and various polymers are being explored, which provide exciting new possibilities. There has also been significant emphasis on developing colloidal nanocrystal-based sensors [27]. Additionally, there is ongoing research in polymeric substrate materials to find the best options for wearable device construction. Commonly used polymeric substrates for wearable devices include versatile options like polydimethylsiloxane (PDMS), high-temperature resistant polyethersulfone, and strong yet flexible polyimide (PI) [28]. Additionally, cost-effective polyethene terephthalate (PET), transparent polyethylene naphthalate (PEN), and polyurethane (PU) are also being explored for their specific advantages. The overall emphasis is on paving the way for a new generation of wearable sensors that are more effective, comfortable, and versatile. In the following sections, various types of PNC-based sensors have been discussed.

Flexible and Responsive: Nanocomposite Temperature Sensors

The advancement of flexible polymer nanocomposite sensors is transforming the field of temperature sensing [29]. Nanocomposite materials are revolutionizing temperature sensors. In comparison to traditional rigid sensors made of pure metals, PNCs-based sensors offer flexibility along with superior sensitivity, faster response times, and improved thermal and electrical properties [7]. PNCs-based temperature sensor sensors have inherent flexibility and are lightweight due to the polymers along with the superior electrical and thermal properties of nanoparticles. Flexible temperature sensors comprise embedding nanofillers, often made from carbon-based materials or metals, within the polymer matrix. There are numerous examples of highly sensitive and responsive temperature sensors. In general, a flexible temperature sensor contains two parts. Primarily, the thermo-sensitive element, whose electrical properties change with temperature fluctuations. Secondly, a flexible substrate that acts as the base layer, typically made of a polymer material, gives flexibility. Additionally, compatible fillers can be incorporated into the sensor at precise concentrations. These fillers can significantly impact the sensor's performance, potentially enhancing its sensitivity, response time, or durability. The combination of different nanofillers and polymer composite generates a conductive pathway, which results in an electrical response to any change in temperature. This kind of sensor offers benefits like increased durability, scratch and wear resistance, flame retardancy, lighter weight, and improved stiffness. Flexible temperature sensors are made by depositing thin films of temperature-sensitive materials or conductive metals like aluminium, silver, gold, or carbon-based materials onto bendable substrates [30]. Carbon-based materials used here include reduced graphene oxide (rGO), graphene and carbon nanotubes (CNTs) [31]. Graphene is a ground-breaking nanomaterial for creating efficient, flexible temperature sensors [14]. The exceptional electrical, physical, thermal, and chemical properties make it an ideal option in sensing applications. Several exceptional properties of graphene, including its large surface area, high electron mobility, superior tensile strength, flexibility, transparency, and biocompatibility, make it a compelling choice for developing stimuli-responsive materials for various sensing applications [14]. In 2006, Stankovich *et al.* published a pioneering work on graphene-based nanocomposites [32]. This electrically conductive material was created by mixing a solution of treated graphene oxide sheets with polystyrene, followed by a chemical reduction process. The polystyrene–graphene composite as reported by the group, exhibits an excellent percolation threshold for room-temperature electrical conductivity, (~ 0.1 S/m), which is sufficient for various electrical applications. Lee *et al.* reported a temperature sensor by fabricating conductive and temperature-sensitive nanosheets made of reduced graphene oxide (r-GO) into a polyurethane (PU) base [33]. This stretchable composite material acts as both the sensor and the flexible layer, allowing it to stretch up to 70% with impressive sensitivity. The sensor's resistance changes by about 1.34% for every degree Celsius temperature variation. Marengo *et al.* fabricated a flexible sensor, which demonstrates both temperature and flow speed sensing [34]. This sensor comprises laser-treated Kapton, as its base and laser-induced graphene as electrodes for sensing. The laser treatment makes a porous structure on the Kapton, which acts as the sensing material. Kapton is chemically made of polyimides. The flow sensor component leverages the pressure-sensitive properties of this porous graphene. A thin layer of PDMS is applied as a protective coating, expanding its application range in both liquids and gases. Lee *et al.* reported a flexible and transparent nanocomposite material by combining rGO with a P(VDF-TrFE) copolymer. This material shows high thermal responsivity in a field-effect transistor, making it suitable for temperature sensor applications [35]. Graphene nanowalls (GNWs) are a unique type of graphene structure. Yang *et al.* fabricated a wearable temperature sensor with exceptional sensitivity composed of GNWs with polydimethylsiloxane (PDMS) [36].

The GNWs film was created on copper foils using a low-pressure radio frequency plasma-enhanced chemical vapour deposition (RF-PECVD) technique. The device exhibits a remarkable positive temperature coefficient of resistivity of 0.214 $°C^{-1}$, which is three times greater than conventional sensors. This significant improvement in sensitivity makes GNWs-based sensors promising for future applications in wearable temperature monitors. High aspect ratio, excellent electrical conductivity, and mechanical strength make carbon nanotubes (CNTs) ideal for stretchable electronics [37]. This has driven research towards CNT-based elastic conductors, exemplified by conductive polymer nanocomposites where CNTs are embedded in a polymer matrix. A flexible temperature sensor using a unique combination of single-walled carbon nanotubes (SWCNTs) and adaptable polymers has been reported recently. The strong covalent bonds formed between the SWCNTs and the polymers boosted the sensor's electrical conductivity. As the temperature rises, segments within the polymers with low glass-transition temperatures detach from the SWCNTs. This allows for a better contact between the nanotubes, ultimately leading to a decrease in the sensor's resistance when heated [37]. In a similar context, Yahav *et al.* demonstrate the potential of CNT-forest-based composite devices to be used as high-performance flexible temperature sensors [38]. These sensors rely on the different ways that CNTs and the surrounding PDMS matrix expand and contract with temperature. This mismatch creates strain within the CNT forest, altering the electrical pathways within the sensor. As a result, the sensor's electrical resistance changes in response to temperature fluctuations. A multi-walled carbon nanotubes (MWCNTs) based flexible and wearable temperature sensor was reported by Nicola *et al.* This sensor incorporates controlled amounts of MWCNTs dispersed within a stretchy plastic matrix made of poly(b-styrene-b-(ethylene-co-but-ne)-b-styrene) [39].

Metals such as gold, silver, platinum, copper, nickel, and aluminium are good conductive fillers due to the availability of free electrons. For example, silver nanowire (AgNW) is a useful conductive material for developing flexible temperature sensors [25]. Cui *et al.* tailored a flexible thermo-sensitive temperature sensor based on AgNW composites in a polyimide thin film. It exhibits a higher temperature coefficient of resistance (TCR) than bulk silver TCR ~3.32x10−3 °C-1 with a sensitivity of 0.47 Ω oC^{-1}. Furthermore, the TCR can be tailored by modifying nanowire density and thermal annealing temperature. This work also reports monitoring the skin temperature at the biceps and knees under motion using this sensor [25]. Overall, flexible temperature sensors have broad applicability in different applications like healthcare monitoring, robotics, and the development of electronic artificial skin.

Flexible Pressure Sensor

The past decade has seen tremendous growth in the development of flexible and stretchable pressure sensors [40]. These sensors can be categorized based on their working principles, *i.e.* piezo-resistivity, capacitance, triboelectricity, and piezoelectricity [40]. Due to their ease of fabrication, these sensors are widely studied. For example, CNTs embedded in a flexible matrix of PDMS have been successful in the advancement of flexible pressure sensors [41]. When pressure is applied to a material containing CNTs dispersed in a polymer, its electrical resistance changes significantly. This phenomenon called the piezoresistive effect, makes these CNT/polymer nanocomposites highly promising for a variety of exciting applications [41]. Gau *et al.* developed a flexible pressure sensor using a spin-coated polyimide (PI) - and MWNT film [42]. The resulting PI-MWNT composite exhibits a highly linear piezoresistive response, making it a promising pressure sensor material. Aerogels made from a composite of reduced graphene oxide and polyimide, created *via* freeze casting and thermal annealing exhibit characteristics of low density, exceptional flexibility, and super-elasticitym [43]. These properties equip sensors with outstanding sensitivity to compression and long-lasting stability.

Li *et al.* demonstrated a piezoresistive thin-film pressure sensor which was fabricated using inkjet printing. The sensor is based on a carbon nanotube-polyimide nanocomposite which enables a ring-shaped design [44]. A flexible CNT/PVA/ZnO composite sensor was reported to be useful for structural health monitoring. The composite sensor showed high piezoresistive sensitivity and stability under bending and impact loading. It is promising for monitoring static and dynamic loads on concrete structures as well [45]. Chen *et al.* reported an electrically conductive elastomer composite [46]. *In-situ* polymerization of pyrrole with a supramolecular polymer matrix is guided by the intermolecular hydrogen bonds between 2-ureido-4[1H]-pyrimidinone (UPy) groups. The elastomer composite with the incorporation of 7.5 wt % polypyrrole (PPy) showed a mechanical strength of 0.72 MPa with an elongation of over 300%, where the rapid self-healing of the mechanical and electrical properties was achieved within 5 min [46].

Conductive polymers are widely used as the primary materials in flexible pressure sensors. Yang *et al.* fabricated pressure sensors based on hierarchically patterned polypyrrole films with a high sensitivity (19.32 kPa^{-1}), a low detection limit (1 Pa), and an ultrafast response (20 ms) [47]. MXene, a 2D layered nanomaterial with high conductivity and a large specific surface area, has been studied for its application in different flexible sensors. For example, Guo *et al.* developed an MXene-based pressure sensor with exceptional sensitivity, flexibility, and degradability [48]. They achieved this by layering porous MXene-infused tissue paper between a thin sheet of biodegradable polylactic acid (PLA) and another thin sheet of PLA coated with interdigitated electrodes. This sensor exhibits high sensitivity with a low detection limit (10.2 Pa), broad range (up to 30 kPa), fast response (11 ms), and low power consumption (10–8 W). Bao *et al.* reported another flexible pressure-sensitive organic polymer transistors with high-pressure sensitivity for application in electronic skin and health monitoring such as the functioning of the heart; with a maximum sensitivity of 8.4 kPa−1, a fast response time of <10 ms, high stability over >15,000 cycles and a low power consumption of <1 mW. This sensor comprises microstructured PDMS dielectric and the semiconducting polyisoindigobithiophene-siloxane in a monolithic transistor [3].

Traditional sensors need bulky, powerful equipment to handle deep-sea pressures, making them unfit for miniaturization and low-power operation. Ocean sensors for soft electronics need to be tough - handling both high pressures in megapascal scale and harsh environments. A new type of sensor, made with ultrathin metal films and encapsulated with PDMS, has been designed for use in the ocean environment. These sensors can tolerate pressure, temperature and salinity changes, making them versatile and suitable for a wide range of applications [49].

Flexible Gas Sensor

Stretchable polymer nanocomposite-based sensors exhibit ultrasensitive gas sensing due to high specific surface area, robust interconnected networks, and effectively exposed sensing materials. This design paves the way for a new generation of flexible, ultrasensitive, stable, and portable chemical vapour sensors for wearable electronics in real-time environmental and healthcare monitoring.

A graphene-based flexible gas sensor in wearable and foldable electronics has been reported recently [50]. This sensor comprises reduced graphene oxide (rGO) and polyaniline in a special layered structure. The sensor is fabricated by polymerizing aniline in a solution containing reduced graphene oxide and then depositing it all onto a flexible PET substrate. This process creates nanoparticles and nanofibers of polyaniline that are anchored to the graphene sheets, forming a network with a high surface area. The resulting gas sensor is transparent, highly sensitive to ammonia gas at low concentrations ranging from 100 ppb to 100 ppm, and has high sensitivity to changes in gas levels. The excellent sensing performance is attributed to the synergistic effects and high surface area (47.896 m2 g−1) of the PPANI/rGO-FPANI network films, efficient artificial neural network sensing channels, and effectively exposed active surfaces. Xue *et al.* fabricated a nanocomposite network by combining PANI nanofibers with CNTs coated in PANI nanoparticles [51]. Ammonium persulfate (APS) was added to a solution containing CNTs and aniline, which triggered the polymerization of PANI and its subsequent deposition as a film. These sensors detect NH_3 at room temperature from 200 ppb to 50 ppm, offering high sensitivity and reliable flexibility.

Bai *et al.* developed healable, transparent electronic gas sensors that work at room temperature based on carbon nanotube network-coated polyelectrolyte multilayers [52]. Pandey *et al.* presented a resistive sensor for detecting ammonia using an ultrasensitive-flexible-silver-nanoparticle-based nanocomposite [53]. The sensor can detect ammonia as low as 500 parts per trillion at room temperature in under a minute. Further, they have established the sensing mechanism. A flexible sensor for detecting hydrogen gas has been reported recently [54]. This sensor is based on a tertiary nanocomposite film of reduced graphene oxide (rGO), tin oxide (SnO_2), and polyvinylidene fluoride (PVDF). A nanocomposite network film is developed by preparing polyaniline/MoS_2 (PANI/MoS_2) nanocomposite directly within a MoS_2 suspension and then depositing it onto a pre-stretched, elastic polydimethylsiloxane substrate. The resulting sensor demonstrates ultrasensitive performance, detecting concentrations as low as 50 ppb [55].

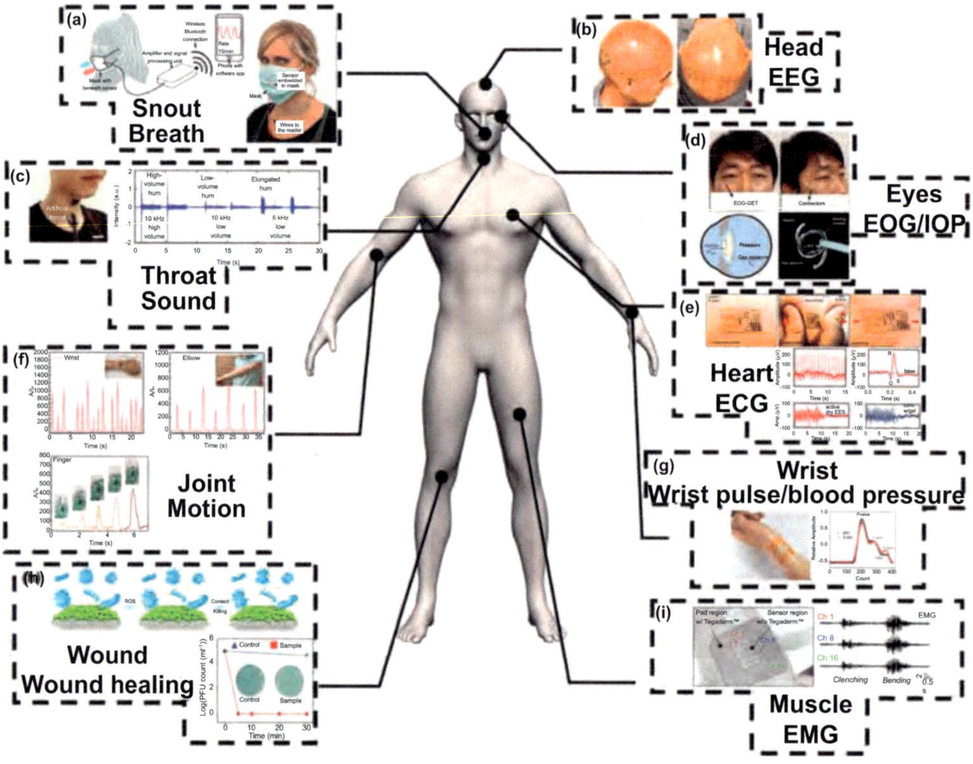

Fig. (3). Wearable health monitors empower portable healthcare, shifting diagnosis from centralized clinics to decentralized, convenient monitoring [61].

Flexible polyaniline/polysiloxane nanocomposite films with tunable properties demonstrate room-temperature NH_3 sensing [56]. Zhang *et al.* have reported a flexible gas sensor based on graphene/ethyl cellulose nanocomposite with a low strain response for volatile organic compounds [57]. Flexible gas sensors for room-temperature formaldehyde (HCHO) detection were fabricated using rGO and rGO/MoS_2 hybrid films on polyethylene naphthalate substrates *via* a simple self-assembly method [58]. A green paper-based flexible room-temperature electronic gas sensor has been reported by Wan *et al.* [59]. The sensor is built using 3D cellulose paper coated with a PANI/rGO nanocomposite network, by *in-situ* PANI polymerization in rGO solution. This flexible sensor shows excellent performance and durability (withstanding 1000 bending cycles), making it suitable for integration into portable electronic devices. The high gas sensing performance is attributed to the high specific surface area (48.532 m^2/g). Apart from gas sensing, a nanocomposite-based material also has applications as humidity sensors [60]. For instance, an armalcolite-PDMS-based flexible capacitive humidity sensor is designed by using the spin coating technique (Fig. **3**), which uses a tiny interdigitated gold (Au) electrode and an ultrathin flexible polyimide base [61].

CONCLUSION

The PNCs have wide and versatile applications in the field of material chemistry. In this chapter, particular emphasis has been given to the fabrication of various flexible sensors based on polymer nanocomposites. Such sensors have been successfully utilised for sensing temperature, and pressure and in other biomedical applications. Owing to excellent mechanical properties, PNC-based sensors have been used in fabricating sensing devices that can be readily used in the form of wearable devices. This design paves the way for a new generation of flexible, ultrasensitive, stable, and portable PNCs. Thus such sensor-based devices have gradually become an essential part of modern lifestyles due to their effectiveness and readiness for live monitoring of health issues. Live and seamless remote monitoring of the functioning of body organs such as the lungs, heart, blood vessels, *etc.* has become possible due to the development of such sensors. Surgical operation on patients remotely from other sides of the world has been made possible due to the development of such sensors, which was an unrealized dream until a few years ago.

Although there is rapid development in the field of flexible sensors, there is still scope for improvement to a great extent. Particularly, it is desirable to have the following characteristics of an ideal sensor: (i) simple operation *i.e.* flawless functioning at ambient conditions; (ii) optimal performance with a low detection limit, high sensitivity, and excellent reproducibility for accurate and reliable data; (iii) fast response to quickly capture rapid changes effectively and (iv) cost-effectiveness and sustainability.

REFERENCES

[1] Abramson, A.; Chan, C.T.; Khan, Y.; Mermin-Bunnell, A.; Matsuhisa, N.; Fong, R.; Shad, R.; Hiesinger, W.; Mallick, P.; Gambhir, S.S.; Bao, Z. A flexible electronic strain sensor for the real-time

monitoring of tumor regression. *Sci. Adv.,* **2022**, *8*(37), eabn6550.
[http://dx.doi.org/10.1126/sciadv.abn6550] [PMID: 36112679]

[2] Someya, T.; Sekitani, T.; Iba, S.; Kato, Y.; Kawaguchi, H.; Sakurai, T. A large-area, flexible pressure sensor matrix with organic field-effect transistors for artificial skin applications. *Proc. Natl. Acad. Sci. USA,* **2004**, *101*(27), 9966-9970.
[http://dx.doi.org/10.1073/pnas.0401918101] [PMID: 15226508]

[3] Schwartz, G.; Tee, B.C.K.; Mei, J.; Appleton, A.L.; Kim, D.H.; Wang, H.; Bao, Z. Flexible polymer transistors with high pressure sensitivity for application in electronic skin and health monitoring. *Nat. Commun.,* **2013**, *4*(1), 1859.
[http://dx.doi.org/10.1038/ncomms2832] [PMID: 23673644]

[4] Sokolov, A.N.; Tee, B.C.K.; Bettinger, C.J.; Tok, J.B.H.; Bao, Z. Chemical and engineering approaches to enable organic field-effect transistors for electronic skin applications. *Acc. Chem. Res.,* **2012**, *45*(3), 361-371.
[http://dx.doi.org/10.1021/ar2001233] [PMID: 21995646]

[5] Wang, X.; Gu, Y.; Xiong, Z.; Cui, Z.; Zhang, T. Silk-molded flexible, ultrasensitive, and highly stable electronic skin for monitoring human physiological signals. *Adv. Mater.,* **2014**, *26*(9), 1336-1342.
[http://dx.doi.org/10.1002/adma.201304248] [PMID: 24347340]

[6] Nyabadza, A.; Vázquez, M.; Coyle, S.; Fitzpatrick, B.; Brabazon, D. Review of materials and fabrication methods for flexible nano and micro-scale physical and chemical property sensors. *Appl. Sci. (Basel),* **2021**, *11*(18), 8563.
[http://dx.doi.org/10.3390/app11188563]

[7] Gao, W.; Ota, H.; Kiriya, D.; Takei, K.; Javey, A. Flexible electronics toward wearable sensing. *Acc. Chem. Res.,* **2019**, *52*(3), 523-533.
[http://dx.doi.org/10.1021/acs.accounts.8b00500] [PMID: 30767497]

[8] Stankovich, S.; Dikin, D.A.; Dommett, G.H.B.; Kohlhaas, K.M.; Zimney, E.J.; Stach, E.A.; Piner, R.D.; Nguyen, S.T.; Ruoff, R.S. Graphene-based composite materials. *Nature,* **2006**, *442*(7100), 282-286.
[http://dx.doi.org/10.1038/nature04969] [PMID: 16855586]

[9] Komarneni, S. Feature article. nanocomposites. *J. Mater. Chem.,* **1992**, *2*(12), 1219-1230.
[http://dx.doi.org/10.1039/jm9920201219]

[10] Manias, E. Stiffer by design. *Nat. Mater.,* **2007**, *6*(1), 9-11.
[http://dx.doi.org/10.1038/nmat1812] [PMID: 17199118]

[11] Stone, D.A.; Hsu, L.; Wheeler, N.R.; Wilusz, E.; Zukas, W.; Wnek, G.E.; Korley, L.T.J. Mechanical enhancement *via* self-assembled nanostructures in polymer nanocomposites. *Soft Matter,* **2011**, *7*(6), 2449-2455.
[http://dx.doi.org/10.1039/c0sm01262a]

[12] Li, Q.; Zhang, L.N.; Tao, X.M.; Ding, X. Review of flexible temperature sensing networks for wearable physiological monitoring. *Adv. Healthc. Mater.,* **2017**, *6*(12), 1601371.
[http://dx.doi.org/10.1002/adhm.201601371] [PMID: 28547895]

[13] Ou, L.X.; Liu, M.Y.; Zhu, L.Y.; Zhang, D.W.; Lu, H.L. Recent progress on flexible room-temperature gas sensors based on metal oxide semiconductor. *Nano-Micro Lett.,* **2022**, *14*(1), 206.
[http://dx.doi.org/10.1007/s40820-022-00956-9] [PMID: 36271065]

[14] He, Q.; Wu, S.; Yin, Z.; Zhang, H. Graphene-based electronic sensors. *Chem. Sci. (Camb.),* **2012**, *3*(6), 1764-1772.
[http://dx.doi.org/10.1039/c2sc20205k]

[15] Stewart, M.E.; Anderton, C.R.; Thompson, L.B.; Maria, J.; Gray, S.K.; Rogers, J.A.; Nuzzo, R.G. Nanostructured plasmonic sensors. *Chem. Rev.,* **2008**, *108*(2), 494-521.
[http://dx.doi.org/10.1021/cr068126n] [PMID: 18229956]

[16] Khan, S.; Ali, S.; Bermak, A. Recent developments in printing flexible and wearable sensing electronics for healthcare applications. *Sensors (Basel),* **2019**, *19*(5), 1230. [http://dx.doi.org/10.3390/s19051230] [PMID: 30862062]

[17] Qin, R.; Hu, M.; Li, X.; Liang, T.; Tan, H.; Liu, J.; Shan, G. A new strategy for the fabrication of a flexible and highly sensitive capacitive pressure sensor. *Microsyst. Nanoeng.,* **2021**, *7*(1), 100. [http://dx.doi.org/10.1038/s41378-021-00327-1] [PMID: 34868631]

[18] da Costa, T.H.; Choi, J.W. Fabrication and patterning methods of flexible sensors using carbon nanomaterials on polymers. *Adv. Intell. Syst.,* **2020**, *2*(5), 1900179. [http://dx.doi.org/10.1002/aisy.201900179]

[19] Wang, X.; Zhang, M.; Zhang, L.; Xu, J.; Xiao, X.; Zhang, X. Inkjet-printed flexible sensors: From function materials, manufacture process, and applications perspective. *Mater. Today Commun.,* **2022**, *31*, 103263. [http://dx.doi.org/10.1016/j.mtcomm.2022.103263]

[20] Tang, N.; Zhou, C.; Xu, L.; Jiang, Y.; Qu, H.; Duan, X. A Fully integrated wireless flexible ammonia sensor fabricated by soft nano-lithography. *ACS Sens.,* **2019**, *4*(3), 726-732. [http://dx.doi.org/10.1021/acssensors.8b01690] [PMID: 30793588]

[21] Wang, Z.; Shaygan, M.; Otto, M.; Schall, D.; Neumaier, D. Flexible hall sensors based on graphene. *Nanoscale,* **2016**, *8*(14), 7683-7687. [http://dx.doi.org/10.1039/C5NR08729E] [PMID: 26997245]

[22] Fruncillo, S.; Su, X.; Liu, H.; Wong, L.S. Lithographic processes for the scalable fabrication of micro- and nanostructures for biochips and biosensors. *ACS Sens.,* **2021**, *6*(6), 2002-2024. [http://dx.doi.org/10.1021/acssensors.0c02704] [PMID: 33829765]

[23] Schroeder, V.; Savagatrup, S.; He, M.; Lin, S.; Swager, T.M. Carbon nanotube chemical sensors. *Chem. Rev.,* **2019**, *119*(1), 599-663. [http://dx.doi.org/10.1021/acs.chemrev.8b00340] [PMID: 30226055]

[24] Hajian, S.; Maddipatla, D.; Narakathu, B.B.; Atashbar, M.Z. MXene-based flexible sensors: A review. *Frontiers in Sensors,* **2022**, *3*, 1006749. [http://dx.doi.org/10.3389/fsens.2022.1006749]

[25] Cui, Z.; Poblete, F.R.; Zhu, Y. Tailoring the temperature coefficient of resistance of silver nanowire nanocomposites and their application as stretchable temperature sensors. *ACS Appl. Mater. Interfaces,* **2019**, *11*(19), 17836-17842. [http://dx.doi.org/10.1021/acsami.9b04045] [PMID: 30985098]

[26] Won, M.; Jung, M.; Kim, J.; Kim, D.S. Fully printed cellulose nanofiber–ag nanoparticle composite for high-performance humidity sensor. *Nanomaterials (Basel),* **2024**, *14*(4), 343. [http://dx.doi.org/10.3390/nano14040343] [PMID: 38392716]

[27] Lorenzon, M.; Pinchetti, V.; Bruni, F.; Bae, W.K.; Meinardi, F.; Klimov, V.I.; Brovelli, S. Single-particle ratiometric pressure sensing based on "double-sensor" colloidal nanocrystals. *Nano Lett.,* **2017**, *17*(2), 1071-1081. [http://dx.doi.org/10.1021/acs.nanolett.6b04577] [PMID: 28032501]

[28] Wu, Y.; Li, Y.; Tao, Y.; Sun, L.; Yu, C. Recent advances in the material design for intelligent wearable devices. *Mater. Chem. Front.,* **2023**, *7*(16), 3278-3297. [http://dx.doi.org/10.1039/D3QM00076A]

[29] Phadkule, S.S.; Sarma, S. Progress in nanocomposite based flexible temperature sensors: A review. Measurement. *Sensors (Basel),* **2023**, *27*, 100692.

[30] Gudadur, K.S.; Veluswamy, P. Flexible reversible polymer nano-composite thin film patch for wearable temperature sensor. *Surf. Interfaces,* **2023**, *39*, 102975. [http://dx.doi.org/10.1016/j.surfin.2023.102975]

[31] Angione, M.D.; Pilolli, R.; Cotrone, S.; Magliulo, M.; Mallardi, A.; Palazzo, G.; Sabbatini, L.; Fine, D.; Dodabalapur, A.; Cioffi, N.; Torsi, L. Carbon based materials for electronic bio-sensing. *Mater. Today,* **2011**, *14*(9), 424-433. [http://dx.doi.org/10.1016/S1369-7021(11)70187-0]

[32] Meyer, J.C., Geim, A.K., Katsnelson, M.I., Novoselov, K.S., Booth, T.J. and Roth, S., The structure of suspended graphene sheets. *Nature,* **2007**, *446*(7131), 60-63.

[33] Trung, T.Q.; Ramasundaram, S.; Hwang, B.U.; Lee, N.E. An all-elastomeric transparent and stretchable temperature sensor for body-attachable wearable electronics. *Adv. Mater.,* **2016**, *28*(3), 502-509. [http://dx.doi.org/10.1002/adma.201504441] [PMID: 26607674]

[34] Marengo, M.; Marinaro, G.; Kosel, J. Temperature and flow sensor from laser-induced graphene. *Proceedings of IEEE Sensors,* **2017**, 1-3. [http://dx.doi.org/10.1109/ICSENS.2017.8234429]

[35] Davaji, B.; Cho, H.D.; Malakoutian, M.; Lee, J.K.; Panin, G.; Kang, T.W.; Lee, C.H. A patterned single layer graphene resistance temperature sensor. *Sci. Rep.,* **2017**, *7*(1), 8811. [http://dx.doi.org/10.1038/s41598-017-08967-y] [PMID: 28821773]

[36] Yang, J.; Wei, D.; Tang, L.; Song, X.; Luo, W.; Chu, J.; Gao, T.; Shi, H.; Du, C. Wearable temperature sensor based on graphene nanowalls. *RSC Advances,* **2015**, *5*(32), 25609-25615. [http://dx.doi.org/10.1039/C5RA00871A]

[37] Yang, H.; Qi, D.; Liu, Z.; Chandran, B.K.; Wang, T.; Yu, J.; Chen, X. Soft thermal sensor with mechanical adaptability. *Adv. Mater.,* **2016**, *28*(41), 9175-9181. [http://dx.doi.org/10.1002/adma.201602994] [PMID: 27572902]

[38] Ben-Shimon, Y.; Ya'akobovitz, A. Flexible and bio-compatible temperature sensors based on carbon nanotube composites. *Measurement,* **2021**, *172*, 108889. [http://dx.doi.org/10.1016/j.measurement.2020.108889]

[39] Calisi, N.; Salvo, P.; Melai, B.; Paoletti, C.; Pucci, A.; Di Francesco, F. Effects of thermal annealing on SEBS/MWCNTs temperature-sensitive nanocomposites for the measurement of skin temperature. *Mater. Chem. Phys.,* **2017**, *186*, 456-461. [http://dx.doi.org/10.1016/j.matchemphys.2016.11.018]

[40] Seesaard, T.; Wongchoosuk, C. Flexible and stretchable pressure sensors: from basic principles to state-of-the-art applications. *Micromachines (Basel),* **2023**, *14*(8), 1638. [http://dx.doi.org/10.3390/mi14081638] [PMID: 37630177]

[41] Xu, M.; Gao, Y.; Yu, G.; Lu, C.; Tan, J.; Xuan, F. Flexible pressure sensor using carbon nanotube-wrapped polydimethylsiloxane microspheres for tactile sensing. *Sens. Actuators A Phys.,* **2018**, *284*, 260-265. [http://dx.doi.org/10.1016/j.sna.2018.10.040]

[42] Gau, C.; Ko, H.S.; Chen, H.T. Piezoresistive characteristics of MWNT nanocomposites and fabrication as a polymer pressure sensor. *Nanotechnology,* **2009**, *20*(18), 185503. [http://dx.doi.org/10.1088/0957-4484/20/18/185503] [PMID: 19420615]

[43] Qin, Y.; Peng, Q.; Ding, Y.; Lin, Z.; Wang, C.; Li, Y.; Xu, F.; Li, J.; Yuan, Y.; He, X.; Li, Y. Lightweight, superelastic, and mechanically flexible graphene/polyimide nanocomposite foam for strain sensor application. *ACS Nano,* **2015**, *9*(9), 8933-8941. [http://dx.doi.org/10.1021/acsnano.5b02781] [PMID: 26301319]

[44] Yang, C.; Li, L.; Zhao, J.; Wang, J.; Xie, J.; Cao, Y.; Xue, M.; Lu, C. Highly sensitive wearable pressure sensors based on three-scale nested wrinkling microstructures of polypyrrole films. *ACS Appl. Mater. Interfaces,* **2018**, *10*(30), 25811-25818. [http://dx.doi.org/10.1021/acsami.8b08666] [PMID: 29993231]

[45] Chen, S.; Luo, J.; Wang, X.; Li, Q.; Zhou, L.; Liu, C.; Feng, C. Fabrication and

piezoresistive/piezoelectric sensing characteristics of carbon nanotube/PVA/nano-ZnO flexible composite. *Sci. Rep.,* **2020**, *10*(1), 8895.
[http://dx.doi.org/10.1038/s41598-020-65771-x] [PMID: 32483263]

[46] Chen, J.; Liu, J.; Thundat, T.; Zeng, H. Polypyrrole-doped conductive supramolecular elastomer with stretchability, rapid self-healing, and adhesive property for flexible electronic sensors. *ACS Appl. Mater. Interfaces,* **2019**, *11*(20), 18720-18729.
[http://dx.doi.org/10.1021/acsami.9b03346] [PMID: 31045346]

[47] Jiang, S., Yu, J., Xiao, Y., Zhu, Y. and Zhang, W., Ultrawide sensing range and highly sensitive flexible pressure sensor based on a percolative thin film with a knoll-like microstructured surface. *ACS applied materials & interfaces,* **2019**, *11*(22), 20500-20508.

[48] Guo, Y.; Zhong, M.; Fang, Z.; Wan, P.; Yu, G. A wearable transient pressure sensor made with mxene nanosheets for sensitive broad-range human–machine interfacing. *Nano Lett.,* **2019**, *19*(2), 1143-1150.
[http://dx.doi.org/10.1021/acs.nanolett.8b04514] [PMID: 30657695]

[49] Li, Y.; Wu, G.; Song, G.; Lu, S.H.; Wang, Z.; Sun, H.; Zhang, Y.; Wang, X. Soft, pressure-tolerant, flexible electronic sensors for sensing under harsh environments. *ACS Sens.,* **2022**, *7*(8), 2400-2409.
[http://dx.doi.org/10.1021/acssensors.2c01059] [PMID: 35952377]

[50] Guo, Y.; Wang, T.; Chen, F.; Sun, X.; Li, X.; Yu, Z.; Wan, P.; Chen, X. Hierarchical graphene–polyaniline nanocomposite films for high-performance flexible electronic gas sensors. *Nanoscale,* **2016**, *8*(23), 12073-12080.
[http://dx.doi.org/10.1039/C6NR02540D] [PMID: 27249547]

[51] Xue, L.; Wang, W.; Guo, Y.; Liu, G.; Wan, P. Flexible polyaniline/carbon nanotube nanocomposite film-based electronic gas sensors. *Sens. Actuators B Chem.,* **2017**, *244*, 47-53.
[http://dx.doi.org/10.1016/j.snb.2016.12.064]

[52] Bai, S.; Sun, C.; Yan, H.; Sun, X.; Zhang, H.; Luo, L.; Lei, X.; Wan, P.; Chen, X. Healable, transparent, room-temperature electronic sensors based on carbon nanotube network-coated polyelectrolyte multilayers. *Small,* **2015**, *11*(43), 5807-5813.
[http://dx.doi.org/10.1002/smll.201502169] [PMID: 26395971]

[53] Pandey, S.; Goswami, G.K.; Nanda, K.K. Nanocomposite based flexible ultrasensitive resistive gas sensor for chemical reactions studies. *Sci. Rep.,* **2013**, *3*(1), 2082.
[http://dx.doi.org/10.1038/srep02082] [PMID: 23803772]

[54] Punetha, D.; Kar, M.; Pandey, S.K. A new type low-cost, flexible and wearable tertiary nanocomposite sensor for room temperature hydrogen gas sensing. *Sci. Rep.,* **2020**, *10*(1), 2151.
[http://dx.doi.org/10.1038/s41598-020-58965-w] [PMID: 32034226]

[55] Yan, H.; Zhong, M.; Lv, Z.; Wan, P.; Yan, H.; Zhong, M.J.; Lv, Z.; Wan, P.B. Stretchable electronic sensors of nanocomposite network films for ultrasensitive chemical vapor sensing. *Small,* **2017**, *13*(41), 1701697.
[http://dx.doi.org/10.1002/smll.201701697] [PMID: 28895272]

[56] Zhang, T.; Cheng, Q.; Jiang, B.; Huang, Y. Design of the novel polyaniline/polysiloxane flexible nanocomposite film and its application in gas sensor. *Compos., Part B Eng.,* **2020**, *196*, 108131.
[http://dx.doi.org/10.1016/j.compositesb.2020.108131]

[57] Zhang, Q.; An, C.; Fan, S.; Shi, S.; Zhang, R.; Zhang, J.; Li, Q.; Zhang, D.; Hu, X.; Liu, J. Flexible gas sensor based on graphene/ethyl cellulose nanocomposite with ultra-low strain response for volatile organic compounds rapid detection. *Nanotechnology,* **2018**, *29*(28), 285501.
[http://dx.doi.org/10.1088/1361-6528/aabf2f] [PMID: 29668484]

[58] Li, X.; Wang, J.; Xie, D.; Xu, J.; Xia, Y.; Li, W.; Xiang, L.; Li, Z.; Xu, S.; Komarneni, S. Flexible room-temperature formaldehyde sensors based on rGO film and rGo/MoS $_2$ hybrid film. *Nanotechnology,* **2017**, *28*(32), 325501.
[http://dx.doi.org/10.1088/1361-6528/aa79e6] [PMID: 28617668]

[59] Yan, H.; Guo, Y.; Lai, S.; Sun, X.; Niu, Z.; Wan, P. Flexible room-temperature gas sensors of nanocomposite network-coated papers. *ChemistrySelect,* **2016**, *1*(11), 2816-2820. [http://dx.doi.org/10.1002/slct.201600648]

[60] Tripathy, A.; Sharma, P.; Sahoo, N.; Pramanik, S.; Abu Osman, N.A. Moisture sensitive inimitable Armalcolite/PDMS flexible sensor: A new entry. *Sens. Actuators B Chem.,* **2018**, *262*, 211-220. [http://dx.doi.org/10.1016/j.snb.2018.01.207]

[61] Deng, Z.; Guo, L.; Chen, X.; Wu, W. Smart wearable systems for health monitoring. *Sensors (Basel),* **2023**, *23*(5), 2479. [http://dx.doi.org/10.3390/s23052479] [PMID: 36904682]

SUBJECT INDEX

A

B

C

D

E

F

G

H

I

L

M

N

O

P

X

Made in the USA
Las Vegas, NV
03 November 2026

57421911R00126